Molecular Biology of Bacteria

Molecular Biology of Bacteria

Edited by
V.V. Tetz and A.A. Totalian

NOVA SCIENCE PUBLISHERS, INC
Commack, NY

Creative Design: Gavin Aghamore
Editorial Production: Susan Boriotti
Assistant Vice President/Art Director: Maria Ester Hawrys
Office Manager: Annette Hellinger
Graphics: Frank Grucci
Acquisitions Editor: Tatiana Shohov
Book Production: Ludmila Kwartiroff, Christine Mathosian, Joanne Metal and Tammy Sauter
Circulation: Iyatunde Abdullah, Sharon Britton, and Cathy DeGregory

Library of Congress Cataloging-in-Publication Data
available upon request

ISBN 1-56072-245-2

6080 Jericho Turnpike, Suite 207
Commack, New York 11725
Tele. 516-499-3103 Fax 516-499-3146
E-Mail: Novascience@earthlink.net

Printed in the United States of America

CONTENTS

List of Tables and Figures

Part 1

Molecular Biology of the Gram-Positive Bacteria

PERSISTENCE OF L-FORM GROUP B STREPTOCOCCUS AND MORPHOLOGICAL CHANGES IN RATE IN RATS HYPOTHALAMUS, PITUITARY AND ADRENAL GLANDS IN EXPERIMENTAL INFECTION

N.D. Vartazarian[1], N.G. Khostikian[1], V.M. Shahmuradian[1], L.G. Gorina[2]

Abstract

We studied morphological changes in the rats' Hypothalamus (H), Pituitary (P) and Adrenal (A) glands after single intraperitoneal administration of a stable L-form Group B Streptococcus (L-GBS) (090 R strain). 2 hours after inoculation of L-GBS, labeled by ^{3}H-thymidine, high indices of the average desintegration of radioactive isotope per minute (dpm) were established in the P, decreasing during 24 hours. In the A with initially low indices, it was observed an acute increase of dpm during 24 hours. In the H dpm was low.

The fact of primary fixation of L-GBS by P at the initial phase of streptococcal bacteriemia was significant and allowed to consider the pathogenetic mechanisms of the development of infectious process from a new point of view.

On this base during a year the prolonged infectious process develops with formation of intratissue streptococcal foci.

Persistence of unlabeled microbs during a year in the H, P and A is accompanied by expressed dystrophic, necrotic and inflammatory changes, having a progressive character.

The disturbance of the hormonal homeostasis is an important condi-

1 Department of Pathological Anatomy, State Medical Institute, Koiun str 2, Yerevan 25, 375025, Armenia

2 Gamaleya Institute of Epidemiology and Microbiology, Academy of Medical Sciences Gamalei 18 Gamales str., Moscow, Russia 123098

tion of maintaining of the prolonged infectious process. Maximally manifesting itself of the early (3,7 and I4 days) and average (I,I.5 and 3 month) terms of the experiment, at the late term (6 and I2 month) it is accompanied by tissue reactions of a compensatory and reparative character.

Infections caused by Group B Streptococcus (GBS) are rather well investigated. These are inflammatory diseases of urogenital tract in women (5,7,II), sepsis and meningitis in newborns (8,IO)as well as widely spread carriage of GBS in genital tracts(5,9).

Concerning L-GBS, its pathogenic potentions, only in very few experimental works the immunologic and morphologic peculiarities of organs' affections in this infection are elucidated (6,I2). Whereas the generalisation of the infectious process with lethal outcome in a part of the patients, or prolonged and flabby course of it, are connected with the response reaction of the organism and biological peculiarities of the microbe and based on the integral function of the connective tissue, endocrine and immune systems.

Taking this fact into consideration the study of the morphologic changes dynamically in the H, P and A in the experiment in infection, caused by L-GBS may help to reveal the mechanisms of progression as well as prolonged course of the pathologic process.

MATERIALS AND METHODS

Animals and L-GBS culture. 90 male albino rats (weight I80-200 g) were inoculated intraperitoneally once with a stable L-GBS (090 R strain) 1.5×10^8 CFU in I ml of physiologic solution. The control group consisted of IO intact animals. The rats were decapitated 3,7 and I4 days (early terms), I,I.5 and 3 months (average terms), 6 months and I year (late terms) after administration.

The Method of Fluid Scintillative Spectrometry and Radioautography

The rats were inoculated by the above described method of labeled ^{3}H-thymidine with L-GBS in the dose 20 mcCi in incubative medium I99 at temperature 37°C during I hr.

For removal of ^{3}H-thymidine which hadn't been included into Streptococci, the labeled microbes were washed by phosphate buffer (pH-6,8) with addition of EDTA.

2,5 and 24 hr after administration of L-GBS the animals were decapitated and all blood was taken. From the H, P and A we took pieces and weighed them.

After homogenisation of the pieces we counted the numbers of decays of radioactive isotope in I ml, accounting for I mg of the mass of each organ by the method of fluid scintillative spectrometry (SL-422I apparatus and "Rosh-Bioelectronic",France).

For radioautographic investigation the serial paraffinic sections from the organs were covered with photoemulsion of "M" type and after exposition stained with hematoxylin and eosin.

Morphometric methods. In paraffinic sections from the Supraoptic (SO) and Paraventricular (PV) nuclei of the H by Gomori were counted IOO neurosecretory cells in IO visual fields of the microscope (x90) in I mm^{2} area.

3 types of cells were distinguished according to their morphometric types.

The I type of cells consisted of large, round and eval cells with a big round nucleus and a well expressed nucleolus. These cells with a high functional activity were devided into 3 subtypes:

a) those with a light cytoplasm, without neurosecretory granules;

b) with a light cytoplasm with insignificant quantity of neurosecretory granules;

c) containing a large quantity of neurosecretory granules

The II type - with albuminous intracellular dystrophy, containing a large quantity of neurosecretory granules.

The III type - shrivelled shapeless cells with a picnotic nucleus and cytoplasm full with neurosecretory granules.

The P and A were weighed after being freed from fatty tissue round of capsule.

In the P the correlation of the cells of adenohypophysis -basophils, acidophils and chromophobes, was defined.

In the A the width of the cortex zone was measured with an ocular micrometer.

The rasults of the morphometric investigations and fluid scintillative spectrometry were statistically worked out by Student.

Histologic and Histochemical Methods.

The H and P were fixed in Buen's liquid and neutral formalin, the A in 96° alcohol and neutral formalin.

Parallel to the ordinary histologic methods, the sections were stained by Gomori for revealence of the neutral fat, neurosecret was determined by Gomori, RNA - by Brachet, DNA - by Feulgen, tigroid substance - by Nissl, PAS reactions were put. The freezed sections of the A were stained with Sudan III, Osmic acid and Nils blue. Streptococci in the sections were stained by Gram and Azure II eosin.

Indirect reactions of Immunofluorescence.

The reactions was put od cryostat sections and imprints from the surfaces of the cut of the studied organ, using rabbit hyperimmune antiserum against the rabbit's globulins (SRI after N.F.Gamalea, Moscow). For the control over the specificity of the fluorescence the sections and imprints of the organs of healthy animals were used, reactions were put with a normal rabbit serum, the method of antigens' blocking by homologous and nonfluorescenting serum was applied.

Transmission Electron Microscopy. Small pieces of H, P and A were fixed in 2.5% glutaraldehyde in phosphate - buffered saline, postfixed in OsO_4. After dehydration in alcohol of increasing concentration the bits embedded in araldid. One - μm thik sections were stained with I% Methylene blue. After inspection of the semithin sections areas of particular interest were selected for ultrastructural analysis.

Ultrathin sections were contrasted with 2% aquous uranyl acetate and viewed a "JSM-30" electrom microscopy.

RESULTS

2 hours after intraperitoneal administration of L-GBS with ^{3}H - thymidine the average decay of radioactive isotope per minute (dpm) was 200.0±I.0 in H (control of the natural radioactive background 28.5±I.2). On the table I one can see that after 5 hr dpm of radioactive isotope remained on the initial level and decreased after 24 hr to I66.2±2.2.

Rather high indices were established in the P, where after 2 hr dpm was I03I.4±I.I, decreasing during 5 hr to 8I0.I±2.3 and after 24 hr - to I60.0±I.I.

In the A, having initially low indices (2I0.2±2.I, 2 hr after inoculation) it was observed acute increase of dpm up to 942.I±0.3 during 24hr.

By radioautographic method in sections the mark of extracellular Streptococci was determined (fig. I).

The investigation of the dynamics of morphologic shifts in the H, P and A in rats during a year revealed following changes.

Hypothalamus. After 24 hr in SO and PV nuclei there were observed mainly cells of Ia and Ib types. On the table one can see that the increase of the quantity of Ib type 2 cells was observed in PV nucleus. In SO nucleus no particular shifts were observed.

The RNA granules were localised perinuclearily and in nervous endings. In electronograms there were revealed shifts in the neuron's nuclei, accumulation of compact chromatin and blocks under the nuclei membranes. There appeared osmiophylic necrotized cells, where the intracellular organelles were found out with difficulty.

Simultaneously there were revealed change in synapses and nervous processes. Sometimes one could observe perivascular and pericellular edema.

Serious changes were revealed beginning from the second week, when large intratissue Streptococcal foci appeared. In SO and PV nuclei the quantity of II and III type increased. Shrivelled and affected cells were found out as well in the posterior H, where side by side with colonies of microbes foci with micronecrosis were observed.

In a number of cases in the H and medullar membranes there were distinguished fresh foci of hemerrhages and perivascular infiltrates of lymphocytes and macrophages.

In the average terms in glial,neurosecretory cells and macrophages L-GBS were found (fig.2). The quantity of II and III type cells was high. The affection was intensified due to the increase of the quantity of neurons with intracellular microbes, development of necrosis foci and plural siderocytes. Sometimes purulent meningoencephalitis was revealed.

There were observed intumescences of meningeal membranes, edema, perivascular cellular infiltrates, microthrombs.

However, at the end of the third month due to elimination the quantity of the intratissue microbal foci significantly decreased.

In the late terms in PV and SO nuclei the percent of the cells with deep changes, dystrophy and necrosis remined high. Degenerative processes were not revealed in the nervous fibers and synapses, pointing to the serious changes in afferent as well as efferent connections of the H . In the shrivelled cells it was observed high concentration of neurosecret (fig.3) . The accumulation of L-GBS was revealed in the cells and out of them.

In all experiments side by side with dystrophy and necrosis there were observed signs of adaptive and regenerative processes. Connected with activation of the neurosecretory cells, there appeared elements with bubble nuclei and big pyroninophilic granulations.

The sizes of the cells of Ia and Ib types reached their maximal quantities. On the table 2 it is shown that the average area of the neurosecretory cells, of the H nuclei significantly increased during a week after inoculation, pertaining till the end of the experiment.

The quantity of hypertrophic neurosecretory cells reached 328.I±2.3 mc^3 (control - 320.I±I.6)

Those processes were closely connected with deep structural shifts in the microcirculatory bed. The vascular changes were strongly expressed round neurosecretory cells, where most of the capillaries were in condition of plethora and stasis. Some of the vessels were narrowed or obliterated by microthrombs.

Pituitary gland. After 24 hr the mass of the P significantly increased and during 2 weeks reached 7.3±I.4 mg (control - 6.8±0.2).

In the adenohypophysis the correlation of basophil, acidophil and chromophobe cells hardly differed from the control quantities. The table 3 shown that the chromophobe cells' quantity prevalied. During 2 weeks the quantity of basophils increased up to I2.8±0.4 (control - 9.I±0.8) with significant decrease of chromophobes up to 47.4±2.3 (control - 49.3±I.4). Nuclei with perinuclear distribution of RNP granules were clearly observed.

By electron microscope under the nuclear membrane of the basophils chromatin became compact. Round the hypertrophic Golgy complex the mitochondria with light or granular matrix and short crysts were distributed.

Widening and swelling of the endoplasmatic reticulum were observed. The quantity of big osmiophilic secretory granules, situated mostly on the periphery of the cell, increased. In sinusoids and neurohypophysis L-GBS were revealed. In the posterior lobe pituicytar cells and Herring's bodies rich with neurosecret were found out.

In the average terms of the experiment on the foreground came disorders of microcirculation, manifested by increase of the vascular permeability, hemorrhage, erythrocytes' sluges. During this period in adenohypophysis the basophils, rich RNP prevailed the normal indices, being on the third month about I4.2±0.4%. The increase of their sizes, quantity and heterogeneity of the secretory granules were observed. Most of the nuclei consisted of compact chromatin, signs of affection of the ultrastructure were observed in cytoplasm. Acidophils formed isolated groups nuclei of which contained from 2 to 4 nucleoli. In some places their cloudy swelling or parenchymatous degeneration were observed. Among chromophobes mytosis were found. In the adenohypophysis accumulations of L - GBS were revealed.

In the posterior lobe along the pituicytar fibers and in the cells granules of neurosecret were determined. Some edema and formation of Herring's bodies were observed too.

The maximal increase of the concentration of basophils in comparison to acidophils was observed a year after inoculation and was equal to I6.2±0.7%. The quantity of big osmiophilic secretory granules, situated mainly on the periphery of the cell, increased too, testifying to the high functional activity. There was decrease of chromophobes up to 45.0± I.4. Parallel to it in the posterior dystrophic changes took place inside the pituicytar cells and edema of fibers developed.

The quantity of Herring's bodies decreased acutely. Thickening of the capsules and coarsening of the gland stroma. were observed.

Adrenal Glands. 24 hr after inoculation all layers of the A cortex were well preserved. Only some activation of the cells zona fasciculata was observed, manifested by the increase of the content of lipid granules and RNP in it. Mitochondria became big, round, with moderately solid bubble - shaped crystalls, sometimes with foci of brightening.

Cells of zona glomerulosa contained plural mitochondria with osmiophilic matrix and peculiar transverse tubular cysts.

After 1-2 weeks there were revealed significant changes in the cortex as well as in the capsule and stroma.

The content of lipids and RNP granules in the cells decreased significantly, small foci of necrosis with presence of L-GBS accumulations were revealed. Thickening and picrinophilia of the fascicles of collagenous fibers were observed. Frequently round the vessels there were distinguished infiltrates of lymphocytes, macrophages and singular plasmatic cells. In the capsule there were foci of mucoid swelling,fibroid changes and small colonies of L-GBS.

In the average terms in the cortex and medullar layers of the adrenal glands necrosis foci were found more frequently.

Vascular shifts, manifested by plethora, stasis, microthrombs with obliteration of the vascular lumen were observed. In definite cases the structure of the cortex was acutely damaged.

In the glomerulosa cells with hyperchromic nuclei and rich with lipids and granules of RNP cytoplasm were determined. Sometimes the glomerules were immersing into zona glomerulosa and without clear borders transferred into zona reticularis with tightly adjacent to each other small cells. The cytoplasm became foamy, vacuoles and lipid inclusions were revealed. Intracellular L-GBS with lysis of cytoplasmatic structures were met (fig.4).

Signs of "tubular dystrophy" were observed, accompanied by accumulations of "empty" bubble - shaped cells with a pressed back to the periphery flat nucleus. These cells were mainly revealed near dilated plethoric capillaries.

On the third month in zona glomerulosa hypertrophic light cells were revealed.In the late terms side by side with dystrophy and necrosis in the cells of zona glomerulosa those changes decreased gradually but were still more than in control.

Simultaneously the regions of hyperplasia and hypertrophy of the cells of zona glomerulosa, forming singular small adenomatous formations in a capsule and subcapsular layer of the A, were determined. Most of the cells of zones glomerulosa were atrophied.

DISCUSSION

The present investigations has demonstrated the localization of intratissue microbal foci and morphologic changes in the rat H, P and A in dynamics during a year, after single intraperitoneal administration of stable L-GBS culture. At the beginning of the experiment we decided to clear of the matter of fixation and localization of L-GBS in the studied organs at the earliest phase of Streptococcal bacteremia.

Our data indicate that during 2 hr the infectious process develops with primary fixation of the microbes in the tissue of P. After 24 hr due to the redistribution of L-GBS in the host organism their concentration decreases, while it increases in the cortex of the A.

The fact of primary fixation of L-GBS in the P in the initial stage of Streptococcal bacteremia, was unexpected and very important, and allowed to consider the pathogenetic mechanisms of the development of the infectious process from a new point of view. The redistribution of L-GBS in the host organism during 24 hr, its primary fixation after the

P in the A cortex are obviously due to the structural - functional peculiarities of the endocrine system organs and character of their interaction with L-GBS.

A very important question appears. What is the cause of the primary fixation of L-GBS in the P at the early phase of Streptococcal bacteremia?

In this process a significant role may have adhesion of microbes in the vessels' endothelium, for it is well known that endothelial cell are the primary targets of microbes, leading to mural thrombus formation, vessel occlusion and mononuclear cell perivascular infiltration (4). Thus the significance of different adhesive mechanisms of virulence of bacteries or local adhesive reactions (3) as well as peculiarities of microcirculation and the degree of vascularization of the organ is not excluded either. But taking into consideration the fact that normal blood serum of animals produces nonspecific blocking effect on Streptococci adhesion (II), unequal quantity of their fixation in the organs is connected with either factors of local and general character.

A definite role may belong to the blood cells, participating in protective and adaptive reactions of the organism at microbe invasion.

It is known, that many types of bacteria aggregate platelets, forming mixed aggregates that may, on the one hand, enhance killing of the bacteria by phagocytes, or, on the other hand, protect the bacteria sequestered in the aggregates from destruction (2,I3)

The results obtained allow to think that the observed shifts in H, P and A reflect natural noncpecific reaction of endocrine organs at the initial phase of the stress and completed by significant structural and functional changes.

Our data testifies damage of the P and following subsequent affection of the A and H, which is very important in the complex chain of the disturbances of endocrine regulation in the infectious process. These initial changes are the main link in the development of pathologic process, influencing the character of the local and general tissue reactions of the organism in persistent Streptococcal infection.

Parallel heavy deviations in the neurosecretory cells of the H on the background of intratissue Streptococci, are reflected on the elaboration of neurohormons and the regulation of the release and inhibition of the anterior pituitary hormons. These changes are complicate due to primary fixation of L-GBS in the P and tissue damages, which lead to the deep hormonal disorders.

The affection of the A by L-GBS and hormonal disorders accompanied with injury in the production of A corticosteroids, which affect many components of the inflammatory and immune response in organism (I).

A definite hormonal disbalance develops, appearing on the base of deep morphologic shifts in the organs of endocrine system. It becames an infavorable background, on which intratissue foci of L-GBS are formed with flabby course of inflammatory reactions.

All following morphologic shifts having a progressive character in conditions of prolonged infection are based on these changes. During a year the prolonged infectious process develops with formation of intratissue Streptococcal foci.

The early and constant sign of the infectious process in the affection of the microcirculatory bed, which having a generalized and progressive character is revealed in the H, P and A. The vascular changes and local tissue reactions are connected with angiotropic effect of the microbe and responce reaction of the host organism.

The prolonged persistence of microbes during a year in the studied organs is accompanied by expressed dystrophic, necrotic and inflammatory changes.

The disturbance of the hormonal homeostasis under the influence of L-GBS and its antigens, is an important condition of maintaining of prolonged infection. Maximally manifesting itself in the early and average terms of the experiment, in the late terms they are accompanied by tissue reactions of compensatory and reparative character. These processes in the late terms are accompanied by partial elimination and decrease of persisting intratissue microbal foci. However, they are not enough for prevention of the prolonged and flabby pathologic process.

Norair Drastamat Vartazarian MD,Ph.D.

Nelli Gurgen Khostikian MD ,PhD

Vartan Maxim Shahmuradian MD

Luisa Georgi Gorina MD,PhD

REFERENCES.

I. Bowen, D.L., and A.S.Fauci. 1988. Adrenal Corticosteroids, p. 877 - 895. In J.I.Gallin,I.M.Goldstein, and R.Snyderman (ed.), Raven Press, Ltd, New York.

2. Clawson, C.C. 1971. Platelet interaction with bacteria. II Fate of the bacteria. Am. J. Pathol. 65: 381 - 398

3. Freter, R., and J.W.Jones. 1983. Models for studing the role of bacterial attachment in virulence and pathogenesis. Rev.Infect.Dis. 5. Suppl.4. 41:647 - 658

4. Jaffe, E.A. 1988. Endothelial Cells: Basic Principles and Clinical Correlates, p.559 - 576. In J.I.Gallin, I.M.Goldstein, and R.Snyderman (ed.),Raven Press, Ltd, New York.

5. Kirmani, N., T.J.Hassan, S.N.Jafarey, and S.Hafiz. 1991. Carriage of group B streptococci in rectum and urogenital tract of pregnant women. J.Pak.Med.Assoc. 41:42 -44

6. Khostikyan, N.G., and N.D.Vartazarian, and L.G.Gorina. 1987. MorphOlogical changes in the neuroendocrine system after inoculation of animals with L-form of streptococcus B. Arch. Patologii. 5:64 - 70

7. Regan, J.A., M.A.Klebanoff, and R.P.Nugent. 1991. The epidemiology of group B streptococcal colonization in pregnancy. Vaginal infections and Prematurity Study Group. Obstet. Gynecol. 77;604 - 610

8. Rubens, C.E., H.V.Raff, J.C.Jackson, E.Y.Chi, J.T.Bielitzki, and S.L.Hillier. 1991. Pathophysiology and histopathology of group B streptococcal sepsis in Macaca nemestrina primates induced after intraamniotie inoculation:evidence for bacterial cellular invasion. J. Infect. Dis. 164: 320,-330

9. Schwartz, B., A.Schuchat, M.J.Oxtoby, S.L.Cochi, A.Hightower, and C.V.Broome. 1991. Invasive group B streptococcal disease in adults. A population-based study in metropoliten Atlanta.JAMA. 266: 1112 -1114

IO. Tessin, I., B.Trollfors, and K.Thiringer.I990. Incidence and etiology of neonatal septicemia and meningitis in Western Sweden. Acta Paediatr. Scand. 79:I023 - I030

II. Totolian, A.A. I990. Streptococcal infections : Streptococcus agalactiae. Streptococcus pyogenes. Vestn. Akad. Med. Nauk SSSR. II: 54 - 60

I2. Vartazaryan, N.D., L.G.Gorina, R.S.Ambardzhanyan, S.A.Goncharova, N.G.Khostikyan, I.V.Zheverzheyeva, M.R.Ter-Kasparova, and E.Kh.Arzakanyan I986. Morphological and Immunological characteristic of organ Damages in Experimental Infection with L-Form of group B streptococcus. Dokl. Akad. Med. Nauk SSSR. 7:22 - 27

I3. Weksler, B.B.I988. Platelets, p. 543 - 557. In J.I.Gallin, I.M.Goldstein, and R,Snyderman(ed),Raven Press, Ltd, New York.

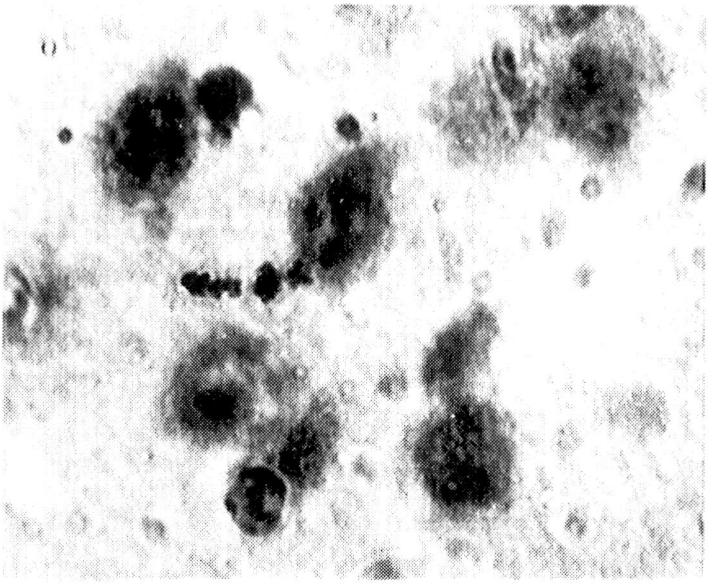

Fig. I

I. Radioautograph of the rats' adenohypophysis 24 hr after intraperitoneal administration of L-GBS, labeled by ^{3}H-thymidine. The group label above extracellular streptococci. Hematoxilin a. Eosin. x900

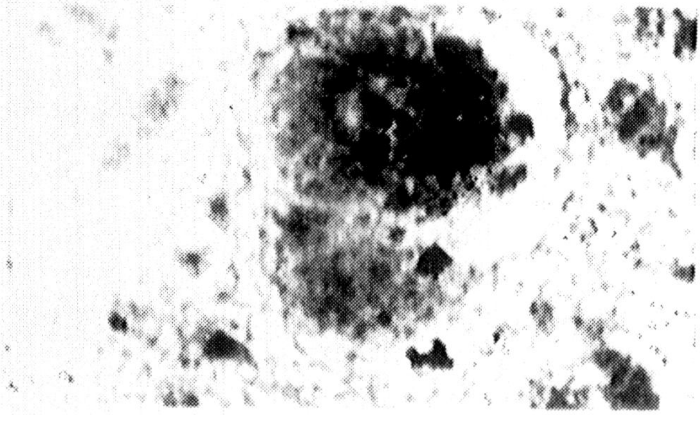

Fig. 2

2. Accumulation of L-GBS in the neurosecretory cell of the H I month after inoculation. Azure II eosin. x900.

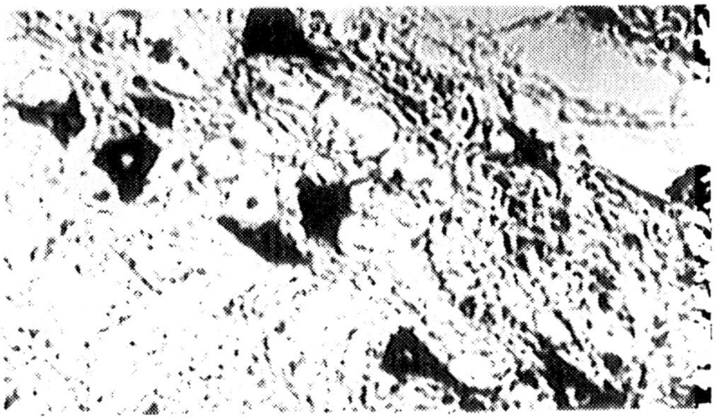

Fig. 3

3. Shrivelled neurosecretory cells of rats' H with high concentration of neurosecret by Gomori, 6 month after inoculation. x400

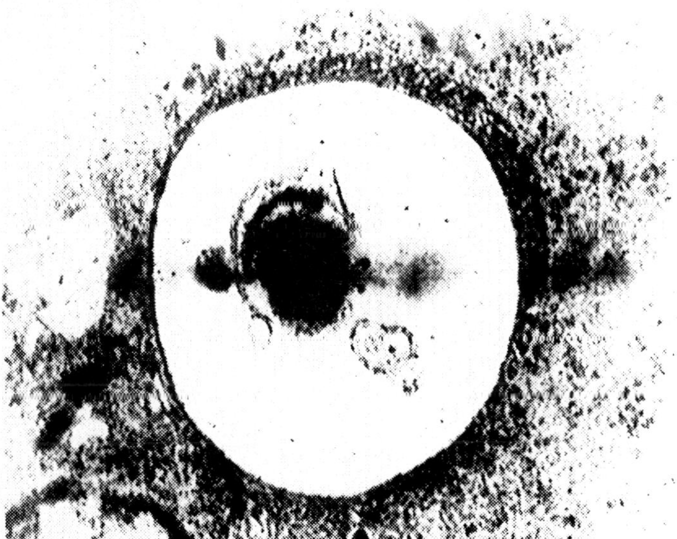

Fig. 4

4. Electronogram of a cell of the fascicular zone of the rats' adrenal cortex 3 month after inoculation with L-GBS. The lysis zone of the cytoplasm around the colony of streptococcus. x29200.

Table I

The mean of dpm of radioactive isotope during 24 hr after single intraperitoneal contamination by a L-GBS, labeled ^3H - thymidine

Organs	Control	Time after contamination 2 hr	5 hr	24 hr
Hypothalamus	28.5±I.2	200.0±I.0	2I8.0±0.5	I66.2±2.2
Pituitary gland	25.I±I.5	I03I.I±I.I	8I0.I±2.3	I60.0±I.I
Adrenal gland	35.3±2.3	2I0.2±2.I	275.2±I.2	942.I±0.3
P < .00I				

Notes. Data give of dpm x I mg tissue $^{-I}$
Data are mean±SD

Table 2

Middle percentage of neurosecretory cells shrivelled forms in SO and PV nuclei of rats' hypothalamus after single intraperitoneal contamination

Time after contamination		PV nucleus Ia	Ib	Ic	2	3	SO nucleus Ia	Ib	Ic	2	3
Control		60	37	3	0	0	36	45	I9	0	0
Early terms	24 hr	57	26	I6	I	0	32	48	I8	2	0
	I wk	50	I6	I6	I7	I	32	4I	I7	4	6
	2 wk	48	20	20	I2	0	30	38	I2	I2	8
Average terms	I mo	40	I6	28	I4	2	32	42	2I	4	I
	3 mo	42	2I	22	I0	5	34	40	I8	5	3
Late terms	6 mo	37	3I	20	8	4	3I	33	2I	I0	5
	I yr	30	35	24	8	3	38	43	I6	2	I

Table 3

Quantitative cytology of basofils, acidofils, and chromophobes in rats' adenohypophysis after single intraperitoneal contamination by L-GBS

Time after contamination		Basophils M±m	Acidophils M±m	Chromophobes M±m
Control		9.I±0.8	4I.4±0.5	49.3±I.4
Early term	24 hr	I0.4±0.4	42.4±0.4	48.I±I.5
	I wk	I2.5±0.8	42.5±0.8	46.2±I.8
	2 wk	I2.8±0.4	44.3±0.6	47.4±2.3
Average term	I mo	I4.6±0.8	40.3±0.4	47.3±I.4
	3 mo	I4.2±0.4	39.2±0.8	46.I±I.3
Late term	6 mo	I5.I±0.8	38.4±0.4	44.3±0.8
	I yr	I6.2±0.7	38.2±0.8	45.0±I.4

A COMPARISON OF CONVENTIONAL TYPING WITH MOLECULAR TYPING OF RECENT ISOLATES OF GROUP B STREPTOCOCCI FROM THE UK 23- 49

A. Efstratiou[*], I.F. Marticorena[*], B.D. Cookson[**]

Abstract

There has recently been an apparent increase in severe invasive disease caused by group B streptococci (GBS) within the UK without any commensurate increase in neonatal or maternal infection. The diseases include endocarditis, septic arthritis, osteomyelitis, meningitis, septicaemia and, strikingly, endophthalmitis. This is reflected by the increase in referrals of adult GBS to our laboratory over a three-year period. By 1992, 25% were isolated from non-

* Respiratory and Systemic Infection Laboratory
** Laboratory of Hospital Infection, Central Public Health Laboratory, London NW9 5HT, UK

pregnancy related systemic diseases in comparison to 10% of referrals in 1989. A subset of 48 recent isolates were chosen in a comparative study of serotyping, phage typing and ribotyping. The isolates covered the spectrum of diseases, 11 different serotypes and 18 phage type patterns. Ribotyping was highly reproducible and typed all isolates into 16 distinct ribotypes using two restriction enzymes *EcoRI* and *HindIII*. When used in a hierarchical manner with serotyping and phage typing, the diversity of these GBS isolates was apparent and resulted in a total of 34 strains. Ribotyping is an effective adjunct in the further subdivision of GBS initially serotyped on the basis of capsular polysaccharide and protein antigens. The recent apparent increase in adult invasive disease does not appear to be related to a restricted number of clones. A second aspect of this study was to examine the distribution of 'new' GBS serotypes amongst non-typables from systemic diseases within the UK over a three year period. Amongst the non-typable GBS (60 of 700), 30% (18/60) belonged to serotype V, 7% to serotype IV and 38% to our new provisional type '3421'. Serotype V was more commonly associated with maternal and neonatal disease with few isolations from severe systemic disease (non-pregnancy related) in adults.

INTRODUCTION.

Group B (GBS) are a major course of morbidity and mortality in neonates, frequently causing septicaemia and meningitis [1,2]. They are also able to cause serious invasive disease in non-pregnant adults [3]. However, reports of adult systemic disease have been somewhat sporadic over the last ten years [3].

Within the UK, there has recently been an apparent increase in severe adult systemic disease without any commensurate increase in neonatal or maternal infection. The types of disease include endocarditis, osteomyelitis, septic arthritis, septicaemia, meningitis and endophthalmitis. The incidence of such infections has not been well documented until recently [4,5]. Schwartz and colleagues (1991) conducted the first population based study to characterise invasive GBS in adults. They emphasised the public health implication of this potentially preventable infection and found a strong association of GBS disease amongst adult populations with chronic illnesses of, for example, diabetes mellitus, chronic soft tissue infections and carcinomas. Meningitis, however, remains a rare disease in older patients. Since 1942, there have been only 42 cases reported [5,6]. GBS adult meningitis was thought to be a disease of patients with chronic debilitating illnesses such as diabetes [5]. A recent review concluded that the incidence of adult GBS meningitis could not be accurately determined but it appears to be a rare disease occurring amongst older individuals, particularly males. Although a large percentage of non-pregnancy related adult invasive disease is noscomially acquired, the reservoir for this organism, the mechanisms for spread and the potential for transmission amongst adults is unknown.

Serotyping is the traditional method used to type GBS. Type specific polysaccharide antigens allow GBS to be classified into serotypes Ia, Ib, II, III, IV, V, provisional 'new' types NT6, M9, 7271. The protein antigens c, R or X are also detectable with polyclonal antisera and supplement the polysaccharide antigen typing scheme. Epidemiological studies of serotype distributions show that GBS disease in neonates and their mothers is generally associated with polysaccharide type III. Whereas, the majority of other infections are distributed amongst the other serotypes [3, 8, 9]. However, serotyping offers low discrimination with a fairly high incidence of non-typability [10].

Bacteriophage typing was developed primarily as an adjunct to serotyping in order to enhance the discriminatory power of serotyping [11]. However, phage typing is also of limited usefulness in that there is a high incidence of non-typability, the method is cumbersome and requires rigid quality control with the use of specific reagents.

Plasmid DNA analysis has been explored and was found not to be generally useful in typing GBS as most isolates did not carry plasmids [12].

Other techniques have been explored, such as multilocus enzyme electrophoresis (MLE),[13] and polyacrylamide gel electrophoresis (SDS-PAGE), [14]. MLE has been used to identify chromosomal genetic diversity and relationships among numerous species and GBS. It was found not to be useful for the differentiation of serotype III isolates as most (particularly from patients with invasive disease) grouped into a single MLE type [13, 15]. SDS-PAGE was found to be useful only as a 'combined technique' with immunoblotting using polyclonal rabbit antisera prepared from adult invasive GBS types [16].

More recently, molecular methods have been assessed by various groups mainly because of the disadvantages and limitations with phenotypic methods.

Gordillo and colleagues (1993) have described the use of pulsed-field gel electrophoresis (PFGE) of restriction endonuclease digests of chromosomal DNA from a collection of GBS isolates [17]. They found that PFGE yielded both reproducible and easily readable patterns. However, they had not fully explored the potential of restriction fragment length polymorphisms (RFLPs) as an epidemiological tool. Restriction fragment length polymorphisms of ribosomal RNA genes (ribotyping) has been described recently by Blumberg and colleagues using a 16S-23S RNA probe derived from *E.coli*. They found high reproducibility and concluded that it was a useful method for subtyping GBS [18]. In the study described here, we performed ribotyping of chromosomal DNA from a collection of GBS isolates from adult invasive disease to assess the potential of this method as a tool for investigating the epidemiology of GBS infections. A comparison was, therefore, undertaken to characterise the subset of GBS isolates by serotyping, phage typing and ribotyping.

In addition, the distribution of new GBS serotypes (ie. types IV, V, NT6, M9, 7271 plus a new antiserum raised within our laboratory, PT3421) was explored as an extension to the present set of antisera. An adult GBS, non-typable isolate was used to raise the 'new' antiserum which was screened against all non-typable isolates referred to our laboratory (both pregnancy related and non-pregnancy related) over a three year period.

1.0 MATERIALS AND METHODS

1.1. Isolates examined

A total of 48 isolates from 42 adult patients were examined in this study (Table 1). The isolates were predominantly from the blood with four exceptions; GBS isolates from knee fluid and vitreous humour were obtained from one case of endophthalmitis and CSF and pus isolates from one case of meningitis.

Sporadic non-related isolates of GBS (60/700) from neonatal, maternal and adult infections referred to our laboratory during a three year period (1990-1992) were screened with the 'newer' GBS type sera.

1.2. Serotyping

Serotyping was performed by standard methods[10]. GBS Lancefield extracts were screened by immunoprecipitation against type specific polyclonal antisera for the polysaccharide types Ia, Ib, II, III, IV, V, the protein antigens c, R and X and a 'new' provisional type serum '3421'. The antisera were prepared within our reference laboratory. Dr J Henrichsen (Statens Seruminstitut, Copenhagen) kindly screened the GBS isolates against the newer types NT6, M9 and 7271.

1.3. Bacteriophage typing

Bacteriophage typing was based on patterns of bacterial lysis by a battery of 31 GBS streptococcal bacteriophages as described [11].

1.4. Ribotyping

Chromosomal DNA was prepared by a modification of the procedures described by Skjold *et al.* [19] and Pitcher *et al.* [20]. A biotin-labelled cDNA probe was prepared from total rRNA from the type strain of *Streptococcus agalactiae* NCTC 8181. It was applied to *HindIII* and *EcoRI* restriction endonuclease fragments as previously described [21]. The rRNA gene restriction patterns were determined as percentage similarities using the Dice coefficient [22]. Briefly, DNA band sizes were determined and Dice coefficients used to compare isolate profiles by scoring positive or negative matches in all loci where there were bands reacting with the cDNA probe. Clustering was based on the unweighted pair group arithmetic average algorithm [23].

2.0 RESULTS

During a three year period (1990-1992), a 50% increase in referrals of GBS isolates from adult invasive disease was observed (Table 2). Prior to 1990 <5% of referrals were from adult systemic disease. During 1990, 55% were from neonatal disease, 35% from maternal disease and only 10% from non-pregnancy related systemic disease in adults. Table 3 shows the predominant serotypes amongst GBS isolates from confirmed infections, isolates from colonisation, carriage and superficial disease were excluded from this study. In 1992, 23% were from adult systemic disease with a decrease in the numbers of GBS from maternal infections. The figures for neonatal disease remain stable. Not surprisingly, the predominant serotypes amongst neonatal infections were III and III/R. However, those more commonly isolated from non-pregnancy related systemic disease belonged predominantly to serotypes Ia/c and Ib/c.

2.1 Serotyping and phage typing

The subset of 48 GBS isolates from adult invasive disease could be subdivided into 11 different serotypes (Ia, Ia/c, Ib, Ib/c, II, II/R, III, III/R, IV/c, PT3421, PT3421/c) and 18 phage type patterns (Table 4). A combination of both schemes resulted in 29 strains. Phage typing was somewhat limited because of its low typability rate, 60% typability (29/48) amongst this subset. The non typability was not restricted to a particular serotype however, 4/4 serotype II/R isolates were non typable by phage as were, 2/7 serotype III, 4/6 serotype III/R and 3/5 GBS isolates belonging to the 'new' provisional type 3421.

2.2 Ribotyping

Unlike serotyping and phage typing, ribotyping offers 100% typability and good reproducibility. The *HindIII* rDNA profiles produced more bands (9-12) than *EcoRI* (5-6). Bands were evenly distributed from 23.1-2.3Kbp for the *HindIII* digests and from 23.1-5.0Kbp for *EcoRI*. Reproducibility was good for all bands. Amongst the rDNA profiles there were many common bands. *HindIII* digests produced 12 ribotype patterns whereas only seven were produced with *EcoRI* (Fig.1). Both enzymes in combination resulted in 16 distinct ribotypes. A total of 29 loci (band positions) were identified. Both restriction enzymes were necessary to produce high discrimination between isolates. Overall, the ribotypes differed by one or more bands. The dendogram (Fig.2) illustrates the average linkage between the 16 ribotypes.

2.3 Serotyping, phage typing and ribotyping

Overall 16 ribotypes were noted among the 48 GBS isolates studied (Table 5). There were four ribotypes amongst the Ia, Ia/c isolates (ribotype pattern numbers 1-4); two of five serotype Ia/c belonged to pattern 1, the remainder were ribotypes 2, 3 and 4. Ribotype pattern number 5 comprised nine isolates belonging to serotypes III (5/9), '3421' (2/9) and IV/c (1/9). Ribotype 6 comprised four isolates, Ib/c (2/4), III (1/4), '3421' (1/4). Ribotype 7 also consisted of a 'large cluster' (15 isolates), this comprised serotypes Ib (4/15), II (1/15), II/R (3/15) and '3421' (1/15). Ribotypes 8-16 comprised single serotypes; for example 5 of 6 GBS of serotype III/R belonged to ribotypes 15 or 16; epidemiologically related isolates always had identical ribotype patterns (unpublished data). Sporadic non-related isolates were quite diverse. This is clearly demonstrated by the GBS type strains (Fig.3) where one or more band differences may be seen.

Phage typing was useful in some instances (despite the low typability rate) as an additional scheme for the subdivision of specific ribotypes; for example, one ribotype pattern, ribotype 7 could be subdivided further into six different groups by phage typing (four isolates were non typable by phage, Table 6). Phage typing in combination with ribotyping resulted in 29 strains. When serotyping, phage typing and ribotyping were used in a hierarchical manner this resulted in a total of 34 strains.

2.4 Distribution of new GBS serotypes within the UK during a three year period (1990-1992)

Non typable GBS (60/700) from neonatal, maternal and adult infections were screened against the 'new' GBS type sera, IV, V, NT6, M9, 7271 and 3421 (a new provisional serotype from the UK derived from a GBS blood culture isolate from an adult patient with endocarditis). The 'new' provisional type 3421 identified 27% (16/60) of the non typables, 30% (18/60) belonged to serotype V, 7% (4/60) to serotype IV and 38% (22/60) were non typable. A small proportion of serotype V also carried protein antigens, for example, V/R (present within three isolates) and V/X (two isolates).

3.0 DISCUSSION

The incidence of non- pregnancy related adult systemic GBS disease has apparently increased within the UK during the last few years. In this study we have attempted to epidemiologically characterise GBS isolates from these diseases using three typing schemes, two based on phenotypic methods (serotyping and phage typing) and molecular typing (ribotyping).

A combination of sero and phage typing resulted in 29 strains from this subset of 48 isolates; however, phage typing was somewhat limited because of its typability rate of 60% (in this study), its rigid quality control requirements using specific reagents that are cumbersome to maintain. Further, the phage patterns were somewhat inconsistent.

More recently DNA-based methods have been described for the characterisation of GBS [18,

24] and have proved to be successful for subtyping GBS further. The results from this study confirm previous findings in that ribotyping offers 100% typability, reproducibility and could differentiate isolates within the same serotype and different serotypes. We have shown that there is genetic heterogeneity within GBS serotypes and that isolates non typable by serotyping or phage typing can be evaluated and compared with other isolates by ribotyping.

When used in a hierarchical manner with serotyping and phage typing, the diversity of these GBS from adult invasive disease was apparent, for example, serotypes Ib/c or III did not belong to a specific ribotype. It was interesting to note the inconsistent correlation between serotype and ribotype. The most common ribotype pattern was produced by GBS of serotypes Ib, Ib/c, II, II/R and PT3421. However, it predominated amongst isolates belonging to serotypes Ib or Ib/c. Our findings demonstrate also that nearly every epidemiologically unrelated strain produced a different ribotype pattern. Whereas, epidemiologically related isolates always had identical ribotype patterns. For our studies, we used a non-radioactive 16+23S probe derived from the GBS type strain which seems to reveal more distinct bands than *E.coli* rRNA probes (unpublished data).

There was no definitive correlation between serotype, phage type or ribotype with a particular disease state.

The extension of the current GBS serotyping scheme with the incorporation of 'new' type sera confirmed the findings of Jelinkova and Motlova [25]. Within the UK, serotype V is predominantly isolated from neonatal and maternal sources and is uncommon among GBS isolated from adult invasive disease. Serotype IV is uncommon within the UK and serotypes NT6, M9, 7271 have, to date not been found amongst UK isolates. The 'new' GBS provisional type 3421 comprised a large proportion of previously non typable GBS isolates. PT3421 could therefore be a candidate for a new serological type of GBS. The geographical distribution of these serotypes needs to be addressed on a wider scale.

The ability of GBS to cause adult invasive disease particularly in patients without any underlying predisposing illness is of concern in view of the fact that GBS septicaemia among adults is apparently increasing with new manifestations of disease such as endophthalmitis as an additional complication [7].

CONCLUSIONS

The findings presented indicate the diversity of GBS from adult invasive disease. Unlike neonatal disease there was no specific association of serotype to disease.

Serotyping in conjunction with ribotyping appeared to be a useful analytical tool to explore the clonal evolution of GBS. Both schemes used in combination were highly successful and reproducible and should therefore prove to be useful in assisting clinical and epidemiological investigations of GBS.

ACKNOWLEDGEMENTS

We wish to thank Dr J.Henrichsen for screening the non typable GBS against NT6, M9, 7271; Dr P.Kuzemenska and Dr J.Jelinkova for kindly providing the new GBS type strains.

REFERENCES

1. R.T. Mayon-White. "The incidence of GBS disease in neonates in different countries". *Antibiotics and Chemotherapy;* Vol.35; 1985; pp.17-27.

2. C.J. Baker and M.S. Edwards. "Group B streptococcal infections". In: *Infectious Diseases of the Foetus and Newborn Infant.* J.S Remington and J.O. Klein (Eds); Philadelphia: W.B. Saunders; 1983; pp.820-881.

3. J.Stringer, J.A. Crees-Morris, S.E.J. Young, R.T. Mayon-White. "Group B Streptococcal systemic disease in Great Britain". *Journal of Infection;* Vol.7; 1981; pp. 251-256.

4. B. Schwartz, A. Schuchat, M.J. Oxtoby, S.L. Cochi, A. Hightower and C.V. Broome. "Invasive group B streptococcal disease in adults". *Journal of the American Medical Association;* Vol. 266; 1991; pp. 1112-1114.

5. R. Sarmiento, F.M. Wilson and R. Khatib. "Group B streptococcal meningitis in adults: case report and review of the literature". *Scandinavian Journal of Infectious Diseases;* Vol. 25; 1993; pp. 1-6.

6. L.A. Rantz and W.M.M. Kirby. "Haemolytic streptococcus bacteremia". A report of thirteen cases with special references to the serologic groups of the etiologic organisms". *New England Journal of Medicine;* Vol. 277; 1942; pp. 730-733.

7. A. Galloway, C.M. Deighton, J. Deady, I.F. Marticorena and A. Efstratiou. "Type V group B streptococcal septicaemia with bilateral endophthalmitis and septic arthritis". *Lancet*; Vol 341; 1993; pp. 960-961.

8. P. Geslin, G. Sissia, J. Jelinkova, A. Fremaux and J. Motlova. "Serotype distribution of group B streptococci from human source in France over a 10-year period 1980-1989)". In: *New Perspectives on Streptococci and Streptococcal Infections*; G.Orefici (Ed.); Stuttgart, Jena, New York: Gustav Fischer. Zentralblatt für Bakteriologie, Supplement 22; 1992; pp. 484-485.

9. S.M. Opal, A. Cross, M. Palmer and R. Almazan. "Group B streptococcal sepsis in adults and infants." *Archives of Internal Medicine*; Vol 148; 1988; pp. 641-645.

10. G. Colman; "Typing of Streptococcus agalactiae (Lancefield Group B)." *European Journal of Clinical Microbiology and Infectious Diseases*; Vol.7; 1988; pp. 226-231.

11. J. Stringer; "Phage typing of Streptococcus agalactiae." In: *Methods in Microbiology*; T. Bergan (Ed); London: Academic Press; 1984; Vol.16; pp.1-22.

12. D.W. Denning, C.J. Baker, N.J. Troup and L.S. Tompkins. "Restriction endonuclease analysis of human and bovine group B streptococci for

epidemiologic study. *Journal of Clinical Microbiology*; Vol.27; 1989; pp.1352-1356.

13. J.M. Musser, S.J. Mattingly, R. Quentin and A. Goudeau. "Identification of a high virulence clone of type III Streptococcus agalactiae causing invasive neonatal disease." *Academic Sciences USA*; Vol.86; 1989; pp.4731-4735.

14. J.A. Elliot, R.R. Facklam and C.B. Richter. "Whole-cell protein patterns of nonhaemolytic group B, Type Ib, streptococci isolated from humans, mice, cattle, frogs and fish". *Journal of Clinical Microbiology*; Vol.28; 1990; pp.628-630.

15. W. Hunter, T. Woods, M. Reeves, A. Schuchat and N. Pigott. "Multilocus enzyme electrophoresis and ribosomal DNA restriction profile analysis of survey isolates of group B streptococcus associated with neonatal infection". In: *Proceedings of the Annual Meeting of the American Society for Microbiology* (Dallas), Washington DC, ASM 1991.

16. I.F. Marticorena, A. Efstratiou, M. Costas and B.D Cookson. "Characterisation of group B streptococci from endophthalmitis and other GBS adult diseases within the UK". In: *Proceedings of the XIIth Lancefield International Symposium on Streptococci and Streptococcal Diseases*, St. Petersburg,

17. M.E. Gordillo, K.V. Singh, C.J. Baker and B.E. Murray. "Typing of group B Streptococci: Comparison of pulsed field gel electrophoresis and conventional electrophoresis". *Journal of Clinical Microbiology*; Vol.31; 1993; pp.1430-1434.

18. H.M. Blumberg, D.S. Stephens, C. Licitra, N. Pigott, R. Facklam, B. Swaminathan and I.K. Wachsmuth. "Molecular epidemiology of group B streptococcal infections: use of a restriction endonuclease analysis of chromosomal DNA and DNA restriction fragment length polymorphisms of ribosomal RNA genes (ribotyping)". *Journal of Infectious Diseases*; Vol.166; 1992' pp.574-579.

19. S.A. Skjold, P.G. Quie, L.A. Fries, M. Barnham and P.P. Cleary. "DNA fingerprinting of Streptococcus zooepidemicus (Lancefield group C an aid to epidemiological study". *Journal of Infectious Diseases*; Vol.155; 1987; pp.1145-1150.

20. D.G. Pitcher, N.A. Saunders and R.J. Owen. "Rapid extraction of bacterial genomic DNA with guanidium thiocyanate". *Letters in Applied Microbiology*; Vol 8; 1989; pp.151-156.

21. A. Efstratiou, G. Colman, D. Crowley and D. Pitcher. "Ribotyping of human pyogenic streptococci". In: *New Perspectives on Streptococci and Streptococcal Infections*. G. Orefici (Ed.); Stuttgart, Jena, New York; Gustav

Fischer; Zentralblatt für Bakteriologie; Supplement 22; 1992; pp.90-91.

22. L.R. Dice. "Measures of the amount of ecologic association between species". *Ecology*; Vol.26; 1945; pp.297-302.

23. M. Costas, R.J. Owen, P.J.H. Jackman. "Classification of Campylobacter Sputorum and allied Camplyobacters based on numerical analysis of electrophoretic protein profiles". *Systematic and Applied Microbiology; Vol.9; 1987;* pp.125-131.

24. Y. Nagano, N. Nagano, S. Takahashi, K. Murono, K. Fujita, F.Taguchi and Y. Okuwaki. "Restriction endonuclease digest patterns of chromosomal DNA from group B β-haemolytic streptococci". *Journal of Medical Microbiology*; Vol.35; 1991; pp297-303.

25. J. Jelinková and J. Motlová. "Worldwide distribution of two new serotypes of group B streptococci: type IV and provisional type V". *Journal of Clinical Microbiology*; Vol.21; 1985; pp.361-362.

Table 1.

Origin of the Isolates Examined

Disease	No of Isolates Examined	Age Range of Patients	No. of Patients
Endocarditis	18	40 - 49 years 50 - 59 years 60 - 69 years 70 - 79 years 80 - 89 years	3 7 5 1 1
Endocarditis + Endophthalmitis	2	60 - 69 years 70 - 79 years	1 1
Endophthalmitis	4	60 - 69 years	2
Meningitis	5	20 - 29 years 60 - 69 years	1 3
Osteomyelitis	3	70 - 79 years 80 - 89 years	2 1
Osteomyelitis + meningitis	2	50 - 59 years	1
Septicaemia	7	70 - 79 years 80 - 89 years	4 3
Septic arthritis	7	50 - 59 years 60 - 69 years 70 - 79 years 80 - 89 years	2 1 2 1

Table 2.

Referrals of GBS isolates from Systemic Disease over a three year period

Year	Adult Invasive Disease	Maternal Infections	Neonatal Infections
1990	19 (10%)	67 (35%)	108 (55%)
1991	84 (28%)	50 (17%)	160 (55%)
1992	55 (23%)	56 (23%)	129 (54%)

Table 3.

Serotype distribution of GBS from Systemic Disease over a three year period

Predominant Serotypes

	1990	1991	1992
Adult	Ib/c, Ia/c, Ib	III, Ia/c, Ib/c	Ib/c, Ib, Ia/c
Maternal	Ia/c, III/R, II/R	Ia/c, III/R, II/R	III/R, Ia/c, NT
Neonatal	III, III/R, Ia/c	III/R, III, Ia/c	III/R, III, Ia/c

Serotype, Phage Type and Ribotype Distribution amongst GBS from Adult Invasive Disease - Table 4

STRAIN NO.	DISEASE	SEROTYPE	PHAGE TYPE PATTERN		RIBOTYPE (HINDIII/EcoRI)	
1	Endocarditis	Ia	NT	**18**	1/1	**1**
2	Endophthalmitis	Ia/C	4/12/14/16/20	**1**	2/1	**2**
3	" + Endocarditis	Ia/C	3/4/7/12/14/16/20	**2**	1/1	**1**
4	Septic Arthritis	Ia/C	NT	**18**	1/1	**1**
5	Endocarditis	Ia/C	NT	**18**	3/2	**3**
6	Septic Arthritis	Ia/C	4/7/12/14/16/20	**3**	4/3	**4**
7	Endocarditis	PT3421	NT	**18**	6/3	**5**
8	Endocarditis	PT3421	7/12	**4**	6/3	**5**
9	Osteomyelitis	PT3421	2/5/7/13	**5**	7/2	**6**
10	Septicaemia	PT3421	NT	**18**	4a/3	**7**
11	Septicaemia	PT3421	NT	**18**	8/3	**8**
12	Endocarditis+	Ib	3/4/12/14/16/20	**6**	4a/3	**7**
13	Endophthalmitis	Ib	3/4/12/14/16/20	**6**	4a/3	**7**
14	Endophthalmitis	Ib	3/4/12/14/16/20	**6**	4a/3	**7**
15	Osteomyelitis	Ib	NT	**18**	5/4	**9**
16	Osteomyelitis	Ib	14	**7**	1/5	**10**
17	Septic Arthritis	Ib	3/4/12/14/20	**8**	4a/3	**7**
18	Septic Arthritis	Ib	3/4/12/14/20	**8**	4a/3	**7**
19	Septicaemia	Ib	3/4/12/14/20	**8**	4a/3	**7**
20	Endophthalmitis	Ib/c	4/12/16/20	**9**	9/6	**11**
21	Endocarditis	Ib/c	12/20	**10**	4a/3	**7**
22	Osteomyelitis + Meningitis	Ib/c	3/4/12/15/20	**11**	4a/3	**7**
23	Osteomyelitis + Meningitis	Ib/c	3/4/12/15/20	**11**	4a/3	**7**
24	Endocarditis	Ib/c	8	**12**	1/5	**10**
25	Endocarditis	Ib/c	NT	**18**	7/2	**6**
26	Septicaemia	Ib/c	3/4/12/14/16/20	**6**	4a/3	**1**
27	Septic Arthritis	Ib/c	NT	**18**	7/2	**6**
28	Septic Arthritis	PT3421/C	5/7/18/11/13/30	**13**	1/1	**1**
29	Meningitis	IV/C	NT	**18**	6/3	**5**
30	Endocarditis	II	12/20	**10**	10/3	**12**
31	Endocarditis	II	12/14/16/20	**14**	4a/3	**7**
32	Endocarditis	II/R	NT	**18**	7/5	**13**
33	Meningitis	II/R	NT	**18**	4a/3	**7**
34	Meningitis	II/R	NT	**18**	4a/3	**7**
35	Meningitis	IIR	NT	**18**	4a/3	**7**
36	Endocarditis	III	7/12	**15**	6/3	**5**
37	Endocarditis	III	7/12	**15**	6/3	**5**
38	Endocarditis	III	NT	**18**	11/7	**14**
39	Endocarditis	III	7/12	**15**	6/3	**5**
40	Meningitis	III	NT	**18**	7/2	**6**
41	Septicaemia	III	8	**12**	6/3	**5**
42	Endocarditis	III	7/12	**15**	6/3	**5**
43	Endocarditis	III/R	NT	**18**	12/3	**15**
44	Septic Arthritis	III/R	12	**16**	6/3	**5**
45	Septicaemia	III/R	NT	**18**	12/3	**15**
46	Endocarditis	III/R	NT	**18**	12/3	**15**
47	Endocarditis	III/R	7	**17**	1/4	**16**
48	Septic Arthritis	III/R	NT	**18**	12/3	**15**

NT: Non-typable   **X**: Pattern number

Table 5.

Distribution of GBS ribotypes from Adult Systemic Disease

Serotype	Ribotype	No. of Isolates
Ia, Ia/c, 3421/c	1 2 3 4	4 1 1 1
Ib/c, III, III/R, 3421, IV/c	5 6	9 4
Ib, Ib/c, II, II/R, 3421	7 8 9 10 11 12 13	15 1 1 2 1 1 1
III, IIIR	14 15 16	1 4 1
Total		48

Table 6.

Use of Phage Typing for the Subdivision of Ribotypes

Ribotype	Phage Type Patterns	No of Isolates within each pattern
	NT	4
		4
7	3/4/12/14/16/20	3
	3/4/12/14/20	1
	12/20	2
	3/4/12/15/20	1
	12/14/16/20	

Fig.1 Schematic representation of the rRNA gene restriction fragments of GBS from adult invasive disease

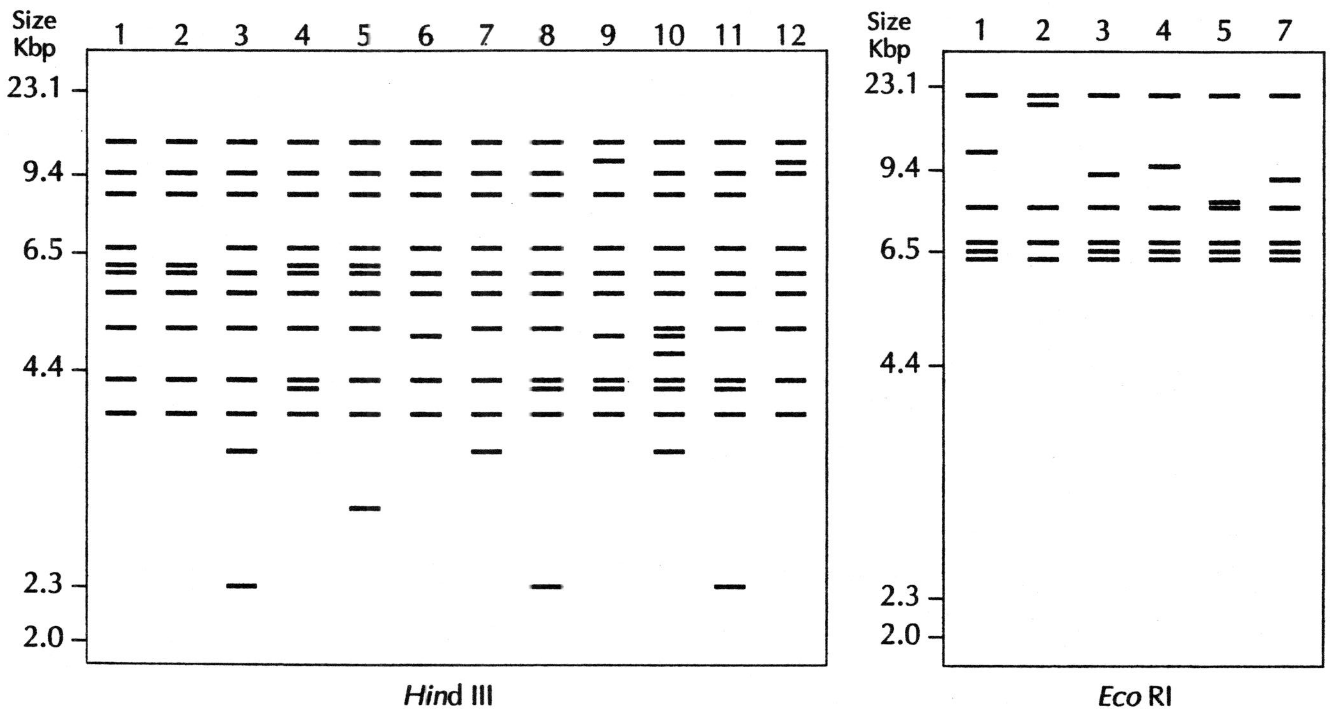

Fig. 2 **Cluster analysis of rRNA genes of GBS from adult invasive disease**

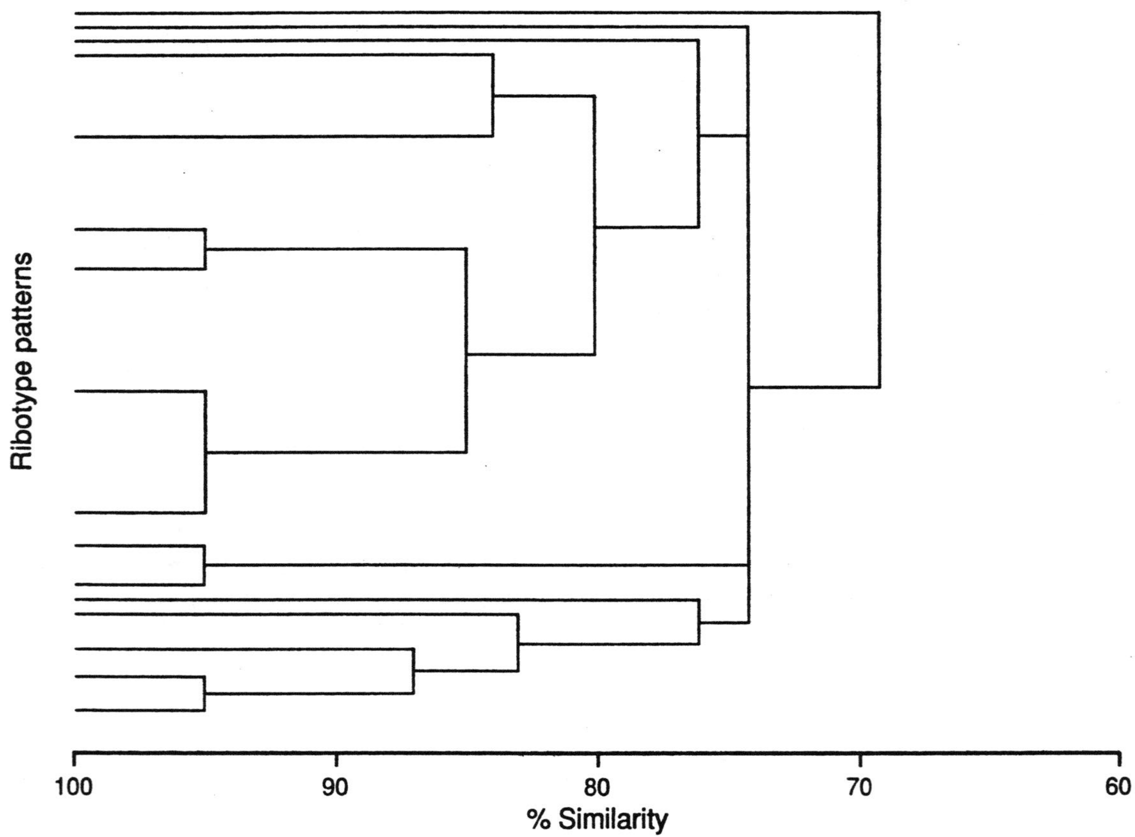

Fig 3. Ribotyping Patterns of the Different GBS Serotypes

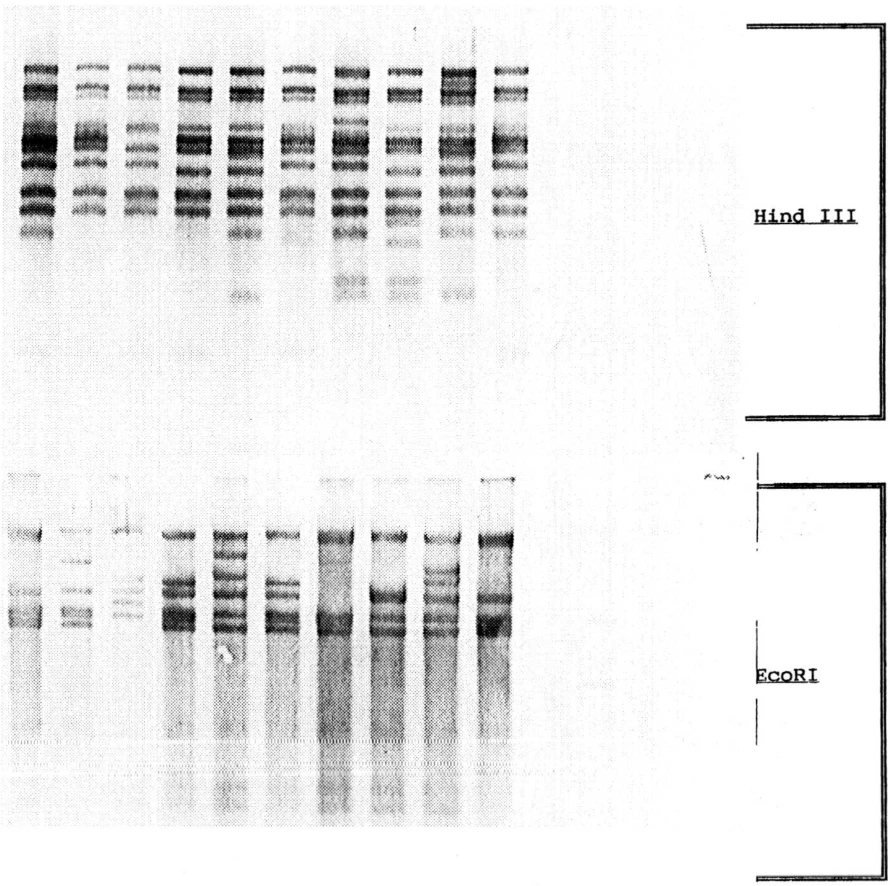

Lanes 1-10: GBS types Ia, Ia/c, Ib/c, II, III, IV, V, R, X, No antigens

Lane 11: Biotin-labelled Hind III Lambda marker (23.1 - 2.0 Kbp)

A CROSS-REACTIVE POLYSACCHARIDE ANTIGEN BETWEEN STREPTOCOCCUS MUTANS AND STREPTOCOCCUS PYOGENES

K. Hirota[1], F. Ota[2], K. Yamashita[3], K. Okada[4], Y. Kuroda[4], H. Inoue[5], C. Uchiyama[5], F. Marchetti[6], E. Ghelardi[6], M. Campa[6] and K. Fukui[1]

Corresponding author: Katsuhiko Hirota, D.D.S., Ph. D.

Department of Oral Microbiology and Immunology,

School of Dentistry, The University of Tokushima,

18-15, 3-chome, kuramoto-cho, Tokushima 770, Japan.

TEL 0886-31-3111 (Japan) EX. 5222

FAX 0886-31-4215 (Japan)

Key words *S. pyogenes*, *S. mutans*, Monoclonal antibody

Epitope, N-acetylglucosamine

---

1 Department of Oral Microbiology and Immunology
2 Department of Food Microbiology
3 Department of Anatomy, School of Dentistry
4 Department of Pediatrics, School of Medicine, The University of Tokushima, Tokushima
5 Kyushu Dental College, Kitakyushu-shi, Kokura, Japan
6 Department of Biomedicine, University of Pisa, Pisa, Italy

ABSTRACT

A monoclonal antibody f-77 (MAb f-77) was prepared by fusing myeloma cells with splenocytes of mice immunized with cells of a *S. mutans* strain SE11. The MAb reacted specifically in enzyme immunoassay (EIA) with whole cells of reference strains of *S. mutans* and *S. pyogenes* and also several isolates of *S. pyogenes* from children with acute glomerulonephritis. It also reacted with carbohydrate and polysaccharide antigen purified from the organisms. Competitive EIA using various haptens revealed that all these immunological reactions were completely inhibited by N-acetylglucosamine. Furthermore, immunoelectron microscopy found that the antigen was on the cell surfaces in both species.

These results indicate that *S. pyogenes* and *S. mutans* share an antigen structure with an epitope of N-acetylglucosamine. It is possible that *S. mutans* may produce the same immune responses in human as does *S. pyogenes* through this immunogenic determinant, and affect an important role in some immunological diseases of streptococcal origin.

INTRODUCTION

Streptococcus pyogenes is known to be one of the main bacterial agents for acute pharyngitis and tonsillitis. It also gives rise to a variety of cutaneous and systemic infections, such as acute rheumatic fever and post streptococcal glomerulonephritis [1]. This microorganism has a complex cell wall structure with various antigenic substances. The group A carbohydrate from *S. pyogenes* consists of rhamnose containing ß-1,3-linked N-acetylglucosamine (GlcNAc) [2], which is responsible for the serologic specificity and induces antibody production [3,4].

S. mutans has been established as the primary causative agent for dental caries but the importance of this bacterial species is not limited to its cariogenicity [5]. It has been implicated in systemic conditions, such as infectious endocarditis [6,7]. The bacterial species has cell surface polysaccharide and the serological classification of this streptococcus is based on the specificity of polysaccharide antigen in its cell surface. It is known that the specific antigens of *S. mutans* (serotypes c, e and f) has rhamnose-glucose polysaccharide covalently bound to the cell wall peptidoglycan [5].

Recently monoclonal antibodies have been used to elucidate immunological properties of mutans streptococci. Among them monoclonal antibody f-77 is specific for *S. mutans* serotypes c, e and f and recognizes a surface polysaccharide antigen with an epitope structure of N-acethylglucosamine [8]. But it has never been demonstrated before that *S. mutans* cross-reacts with *S. pyogenes* isolated from patients with gromerulonephritis. The purpose of this present study is to demonstrate immunological cross-reactions between the two bacterial species using enzyme immnoassay (EIA), competive EIA and immunoelectron microscopy.

MATERIALS AND METHODS

1. Bacterial strains and Culture

A refrence strain of *S. pyogenes* ATCC 19615, A374 (Strain A374 was obtained from the Rockefeller University culture collection) and several strains of the species isolated from children with acute glomerulonephritis (TDP-1, 7y, female; TDP-2, 9y, male; TDP-3, 5y, male; TDP-4, 8y, female) were used. The mutans streptococci used included *S. cricetus* HS1, E49 (serotype a), *S. rattus* FAI, BHT (serotype b), *S. mutans* Ingbritt, NCTC10449 (serotype c), LM7, P4 (serotype e), SE11, OMZ175 (serotype f), *S. sobrinus* OMZ176, MT615R (serotype d), OMZ65、 6715 (serotype g), *S. downei* MF25, BFe12 (serotype h). As previously described [9], bacterial cells were cultured in Brain Heart Infusion broth (BHI, Difco Laborat., Detroit, U.S.A.) supplemented with 0.5%(w/v) glucose for 18h at 37°C.

2. Carbohydrate antigen of *S. pyogenes* and Polysaccharide antigen of *S. mutans*

The carbohydrate antigen of *S. pyogenes* [10] was a generous gift from Prof. Tosizi Ohkuni, Nippon Medical School, Tokyo, Japan. This consists of rhamnose and N-acctylglucosamine in a ratio of approximetely 2 : 1 and used for EIA at a concentration of 1.6 μg/50μl in distilled water. Polysaccharide antigens of *S. mutans* were purified on DEAE-Sephadex A-25, Sephadex G-200 and CM-Sephadex A-25 columns from cultured cells of *S. mutans* Ingbritt, LM7, SE11 as described previously [8]. These were used for EIA at a concentration of 1.0 μg/50μl in distilled water.

3. Preparation of MAb f-77

Preparation of MAb f-77 has been described previously [8]. Briefly, BALB/c mice were immunized intraperitoneally with cells of *S. mutans* strain SE11, emulsified with Freund's complete adjuvant. A few weeks later, the mice were reimmunized with the same type of cells in Freund's incomplete adjuvant. After four weeks, a bacterial suspension in PBS of the same serotype was given intravenously. Three days later their splenocytes were prepared and fused with myeloma cells (Sp2/0-Ag14) in the presence of 50%(w/v) poly-ethyleneglycol 4000 (Merck, Germany). Fused cells were cultured and antibody-secreting cells were selected. The desired hybridomas were cloned at least twice by the limitig dilution method. MAb f-77, which was found to be of the immunoglobulin M class, was obtained from ascitic fluids of pristane-primed BALB/c mice. A 50% ammonium sulfate-precipitable fraction was used for analysis.

4. Enzyme immnoassay

MAb f-77 was tested for specificity following the method of Ota, et.

al. [9]. Briefly, bacterial strains were cultured, harvested, and washed according to the method described above. Bacterial suspensions were made in phosphate-buffered saline (PBS, pH 7.5) to give an O.D. reading of 0.6 at 660nm. Fifty µl of the cell suspension were dispensed into immunoplate wells (disposable sterile polystyrene EIA plate 25801, Corning Glass Works, N.Y., U.S.A.) and the plates were allowed to dry completely by incubation at 37°C overnight. The wells were coated with 1.0% bovine serum albumin in PBS (BSA-PBS). Then the immunoplate wells were treated with 50 µl of MAb f-77(x1000) at room temperature for 60min, washed with PBS, and then reacted with 50µl of anti-mouse IgM labelled with horseradish peroxidase at room temperature for 60min (x2000). Finally the wells were washed with PBS and reacted with a substrate mixture of H_2O_2 and ABTS (Kirkegaard and Perry Labs., Gaithersburg, U.S.A.) for 60min at room temperature. Absorbance was measured for each well at 414nm in an immunoreader (EIA reader, model 2550, BioRad Lab., U.S.A.). Controls included wells reacted with labelled antibody and a substrate mixture of H_2O_2 and ABTS alone.

5. Competive EIA

A competive EIA was performed in microtiter plate wells coated with whole cells or purified antigen of *S. pyogenes* or *S. mutans* as described previously [9]. Briefly, after treatment with BSA-PBS, the immunoplate wells were first treated with a mixture of MAb f-77(x1000) and various amounts of an inhibitor. The mixture had been reacted for 60m in a microcentrifuge tube prior to addition to the wells. Then the immunoplate wells were treated in the same manner for the ordinary EIA procedures described above. Inhibition was expressed as percentages of the EIA values relative to those obtained in the absence of inhibitors.

6. Immunoelectron microscopy

Culture cells of *S. mutans* strain Ingbritt and *S. pyogenes* strain ATCC 19615 were washed in PBS, and cells were treated with 1% BSA-PBS at room temperature for 30 min. They were reacted with MAb f-77(x1000) at room temperature for 60min. After washing they were treated at room temperature for 60min with goat anti-mouse IgM conjugated to colloidal gold (Amersham, Lab., U.K.) diluted in 1:20. Contrc samples involved treatment with goat anti-mouse IgM conjugated to colloidal gold alone. For electron microscopic observation, treated cells were fixed for 2 h at 4°C. They were resuspended in PBS and mixed with an equal volume of 0.5% uranyl acetate. Immediately afterwards, a drop of the mixture was placed on a 200-mesh grid coated with a formvar membrane, and observed in a transmission electron microscope (H-800, Hitachi, Japan).

RESULTS

Reactivities of MAb f-77

Reactivities of MAb f-77 were examined against whole cells of *S. pyogenes* and several mutans groups of streptococci. MAb f-77 reacted specifically with all strains of *S. pyogenes* (Table 1). However It reacted only with *S. mutans* Ingbritt, NCTC10449, LM7, P4, SE11, OMZ175 and not with cells of the other serotype strains tested. As previously shown, MAb f-77 reacted specifically with polysaccharide antigens purified from *S. mutans* serotypes c, e, and f and not with cells of the other serotype strains tested (Table 2). MAb f-77 reacted strongly with purified carbohydrate antigen of *S. pyogenes* A374. A similar reactivity was observed between *S. pyogenes* and *S. mutans* in terms of antigen quantity.

Inhibitory effects of haptenic sugar and sugar derivaties on the reactivity of MAb f-77 were examined by EIA using whole cells of *S. pyogenes* A374, and three strains of *S. mutans* Ingbritt, LM7, OMZ175. A strong inhibition was demonstrated with N-acetylglucosamine at a concentration of 15 µmol/50µl, whereas the other haptens did not show substantial inhibitory effects at that concentration (Table 3). When MAb f-77 was tested against purified polysaccharide antigens of *S. pyogenes* and *S. mutans* in the presence of various inhibitors, N-acetylglucosamine showed the strongest inhibition. The other inhibitors did not show any substantial effects in a range of 1.0-5.0 µg/well.

Figure 1 shows negatively stained immunoelectron micrographs of *S. pyogenes* strain ATCC19615 and *S. mutans* strain Ingbritt. Cells were reacted with MAb f-77 and gold-labelled anti-mouse IgM. Many gold particles were observed diffusely on the cell surfaces in both species. Bar represent 0.1 µm.

DISCUSSION

MAb f-77 reacted in enzyme immunoassay with whole cells of a reference strain (ATCC19615, A374) and clinical isolates of *S. pyogenes* from children with acute glomerulonephritis. Previously it was demonstrated that the immunological reaction between the MAb f-77 and a polysaccharide antigen of *S. mutans* was due to an antigenic determinant having a chemical structure of N-acetylglucosamine. It also reacted with polysaccharide antigen purified from 3 serotypes of *S.mutans*, as well as carbohydrate antigen from *S. pyogenes*. However it did not react with many strains of oral streptococci such as *S. sanguis* ATCC10556, *S. parasanguis* ATCC15912, *S. oralis* ATCC10557, *S. mitis* ATCC15910 (Data not shown). Thc present study clealy demonstrated that *S. mutans* has an antigen structure on its cell surface with an epitope of N-acetylglucosamine in common with *S. pyogenes*.

It has been reported that patients with acute rheumatic fever and acute glomerulonephritis showed a significantly high level of anti-N-acetylglucosamine antibody in their serum [10]. A serum obtained from a patient with acute rheumatic fever reacted with whole cells of *S.*

mutans NCTC10449 [11]. Indeed, glomerulonephritis and intestitial nephritis were reproducibly induced in rabbits by injection of *S. mutans* cells, the lesions observed in the kidneys were similar, in some aspects, to poststreptococcal glomerulonephritis in man [12]. It has been known for many years that *S. mutans*, a species of the mutans group is deeply associated with the induction of dental caries. The presence of *S. mutans* has development of dental caries in childhood. High antibody titres against *S. mutans* may have a role in protection against dental caries [13]. but the effects of much high titres in systemic disease is not fully understood. It is well known that whole cells of *S. pyogenes* induce an inflammatory response as the main physiological stimulus for the production of tumor necrosis factor-α (TNF-α) and interleukin 1-ß (IL-1ß) by monocytes [14]. The serotype polysaccharide from *S. mutans* can also induce the TNF-a and IL-1ß [15].

Although the present study demonsrates obvious cross-reaction between *S.pyogenes* and *S. mutans in vivo,* through an epitope of N-acetylglucosamine, it is unknown how such immune responses by *S. mutans* will influence the initiation and outcome of streptococcal infections. Further studies are awaited to clarify these aspects.

ACKNOWLEDGEMENTS

This study was supported by Grants-in-Aids of Science and Culture from the Ministry of Education of Japan (No 04671266, 04044130)

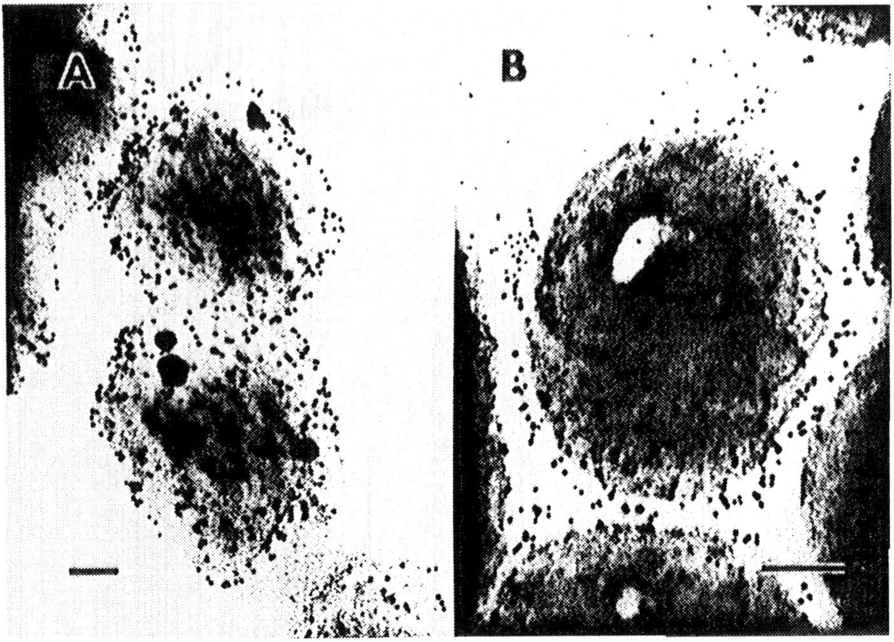

FIGURE LEGEND

Figure 1. Negatively stained Immunoelectron micrographs of *S. pyogenes* ATCC19615 (A) and *S. mutans* Ingbritt (B). Cells were reacted with MAb f-77 and gold-labelled anti-mouse IgM before they were processed for electron microscopy. Note the gold particles on the cell surfaces. Bars represent 1.0 μm.

Table 1 **Reactivity of monoclonal antibody f-77 with whole cells of *S. pyogenes* and mutans streptococcal strains as determined by enzyme immunoassay.**

Strain	type	Reactivity (Means± SD)
S. pyogenes		
ATCC19615	—	102.20±1.40*
A374	—	104.27±1.04*
TDP-1	—	102.52±1.04*
TDP-2	—	103.32±0.78*
TDP-3	—	102.80±0.54*
TDP-4	—	104.65±1.05*
S. cricetus		
HS1	(a)	1.80±0.38
E49	(a)	1.77±0.20
S. rattus		
FA1	(b)	2.32±0.18
BHT	(b)	3.12±0.08
S. mutans		
Ingbritt	(c)	100.00±0.00*
NCTC10449	(c)	101.52±0.86*
LM7	(e)	99.50±1.64*
P4	(e)	99.52±0.62*
SE11	(f)	99.35±1.31*
OMZ175	(f)	99.05±1.13*
S. sobrinus		
OMZ176	(d)	1.80±0.26
MT615R	(d)	2.05±0.13
OMZ65	(g)	2.20±0.38
6715	(g)	2.62±0.26
S. downei		
MF25	(h)	1.92±0.18
BFe12	(h)	1.80±0.34

Values are expressed as % absorbance at 414 nm relative to that with *S. mutans* strain Ingbritt.

*, Statistically significant (P <0.001) from results obtained with negative control.

—, Not applicable.

Table 2 **Reactivity of monoclonal antibody f-77 with carbohydrate (CA) antigen of *S. pyogenes* and polysaccharide (PS) antigen of mutans streptococcal strains as determined by enzyme immunoassay.**

CA or PS antigen from	type	Reactivity (Mean±SD)
S. pyogenes		
A374	—	108.47±2.66*
S. cricetus		
HS1	(a)	1.60±0.26
S. rattus		
FA1	(b)	3.70±0.52
S. mutans		
Ingbritt	(c)	100.00±0.00*
LM7	(e)	99.47±1.26*
SE11	(f)	98.67±0.86*
S. sobrinus		
MT615R	(d)	1.30±0.20

Values are expressed as % absorbance at 414 nm relative to that with *S. mutans* strain Ingbritt.

*, Statistically significant ($P < 0.001$) from results obtained with negative control.

—, Not applicable.

Table 3 Inhibitory effects in enzyme immunoassay of various haptenic sugars and their derivatives on the reactivity of monoclonal antibody f-77 with whole cells of *S. pyogenes* and *S. mutans*

Haptens (15μ mol/well)	Inhibition (Mean±SD)			
	S. pyogenes A374	*S. mutans* Ingbritt	*S. mutans* LM7	*S. mutans* SE11
D-Glucose	1.10±0.08	1.32±0.28	1.25±0.35	1.87±0.44
D-Galactose	1.37±0.42	1.35±0.33	1.67±1.76	1.55±0.93
D-Mannnose	1.05±0.05	1.02±0.68	1.42±0.92	1.20±0.38
L-Rhamnose	1.80±0.26	1.37±0.42	1.87±0.78	1.27±1.04
D-Glucosamine	1.35±0.13	1.27±0.38	1.32±0.42	1.60±0.42
D-Galactosamine	1.42±0.08	1.52±0.50	1.80±1.78	1.97±1.08
D-Mannosamine	1.55±0.13	1.75±0.61	1.50±0.86	1.70±0.30
N-Acetylglucosamine	100.00.±5.46*	100.0±0.00*	100.0±5.34*	100.00±4.68*
N-Acetylgalactosamine	3.27±3.98	4.02±3.00	4.90±2.62	2.72±4.56
N-Acetylmannosamine	1.25±1.61	1.50±0.98	1.50±0.98	1.40±0.62
Cellobiose	1.30±0.38	1.30±0.38	1.60±0.54	1.25±0.25
Gentibiose	1.92±0.76	1.60±0.98	1.57±1.28	1.47±0.52
Lactose	1.30±0.38	1.00±0.66	1.90±0.34	1.22±0.32
Maltose	1.17±1.22	1.72±0.36	1.32±0.22	1.85±1.53
Melibiose	1.25±2.01	1.22±0.32	1.75±0.21	1.27±0.26
Raffinose	1.42±0.62	1.46±0.42	1.40±0.46	1.52±1.50

Values are expressed as % absorbance at 414 nm relative to that with *S. mutans* strain Ingbritt

* , Statistically significant (P <0.001) from results obtained with negative control.

Table 4 **Inhibitory effects in enzyme immunoassay of various haptenic sugars and their derivatives on the reactivity of monoclonal antibody f-77 with carbohydrate antigen (CA) of *S. pyogenes* and polysaccharide antigen (PS) of *S. mutans***

Haptens (15 μ mol/well)	Inhibition (Mean±SD)	
	S. pyogenes A374(CA)	*S. mutans* Ingbritt (PS)
D-Glucose	1.12±1.12	1.45±0.81
Galactose	1.35±0.69	1.37±0.62
D-Mannnose	1.20±0.90	1.52±0.50
L-Rhamnose	1.35±0.45	1.20±0.26
D-Glucosamine	1.20±0.30	1.55±0.91
D-Galactosamine	1.12±0.86	1.20±0.26
D-Mannosamine	1.55±1.61	1.47±1.08
N-Acetylglucosamine	100.00±5.38*	100.00±0.00*
N-Acetylgalactosamine	4.75±2.41	3.72±1.42
N-Acetylmannosamine	1.90±0.74	1.57±1.58
Cellobiose	1.50±0.72	1.42±0.92
Gentiobiose	1.42±0.64	1.42±1.00
Lactose	1.50±0.86	1.30±0.36
Maltose	1.40±0.52	1.47±1.02
Melibiose	1.50±0.74	1.62±1.16
Raffinose	1.42±0.44	1.15±0.21

Values are expressed as % absorbance at 414 nm relative to that with *S. mutans* strain Ingbritt.

* , Statistically significant ($P < 0.001$) from results obtained with negative control.

REFERENCES

1 Dudding, B. A. and Ayoub, E. M., Persisitence of streptococcal group A antibody in patients with rheumatic valvular disease. *J. Exp. Med.*, 128, 1968, p. 1081.

2 Krause, R. M., Symposium on the relationship of structure of microrganisms to their immunological properties. IV. Antigenic and biochemical composition of hemolytic Streptococcal cell walls. *Bacteriol Rev*, 27, 1963, p. 369.

3 Goldstein, I., Halpern, B. and Robert, L., Immunological relationship between Streptococcus A polysaccharide and the structural glycoproteins of heart valve. *Nature*, 213, 1967, p. 44.

4 Goldstein, I., Rebeyrotte, P., Parleba, J. and Halpern, B., Isolation from heart valves of glycopeptides which share immunological properties with Streptococcus hemolyticus group A polysaccharides. *Nature*, 219, 1968, p. 866.

5 Hamada, S. and Slade, H. D., Biology, immunology and cariogenicity of *Streptococcus mutans*. *Microbiol.Rev.*, 44, 1980, p. 331.

6 Harder, E. J., Wilkowski, C. J., Washington, A. J. and Geraci, J. E., *Streptococcus mutans* endocarditis. *Ann. Intern. Med.*, 80, 1974, p. 364.

7 Lockwood, W. R., Lawson, L. A., Smith, D.L., McNeill, K. M. and Morrison, F. S., *Streptococcus mutans* endocarditis : Report of a case. *Ann.Intern.Med.*, 80, 1974, p. 369.

8 Nagamune, H., Ota, F., Senesi, S., Batoni, G., and Fukui, K. Monoclonal antibodies specific for a polysaccharide antigenic determinant of *Streptococcus mutans* (Serotype c, e and f). In: New Perspectives on Streptococci and Streptococcal Infections (Orefici, G Ed), Gustav Fischer Verlag, Stuttgart. 22, 1992, pp. 223-225

9 Ota, F., Hirota, K., Nagamune,H.,Senesi,S., Satomi, Y., Yasuoka, M., Akiyama, Y., Fukui, K.and Kawata,T., A comparison of competitive enzyme immunoassay and precipitin inhibition tests in the analysis of polysaccharide antigenic determinants of oral streptococci. *J. Immunol. Methods.*,124, 1989, p. 259.

10 Todome, Y., Ohkuni, H.,Yokomuro, K.,et al. Enzyme-linked immunosorbent assay of antibody to group A Strptococcus specific C carbohydrate with trypsin-pronase-treated whole cell as antigen. *J Clinical Microbiol*, 26, 1988, p. 464.

[11]. Van de Rinji, Bleiweis, A.S. and Zabriskie, J.B. Antigen in *Streptococcus mutans* cross reactive with human heart muscle. *J. Dent. Res.*, 55, 1976, p. C59.

[12]. Stinson, M. W., Nisengard, R. J., Neiders, M. E., and Albini, B. Serology and tissue lesions in rabbits immunized with *Streptococcus mutans. J. Immunology*, 131, 1983, p. 3021.

[13]. Lehner, T., Murray, J. J., Winter, G. B., Caldwell, J. Antibodies to *Streptococcus mutans* and immunoglobulin levels in children with dental caries. *Archs oral. Biol.*, 23, 1978, p. 1061.

[14]. Yamamoto, A., Usami, H., et al.: The use of lipoteichoic acid(LTA) from *Streptococcus pyogenes* to induce a serum factor causing tumor necrosis. *Br J Cancer*, 51,1985, p. 739.

[15]. Benabdelmoumene, S., Dumont, S., Petit, C., Poindron, P, Wachsmann, D. and Klein, J.P.: Activation of human monocytes by Streptococcus mutans serotype f polysaccharide: Immunoglobulin G Fc receptor expression and tumor necrosis factor and interleukin-1 production. *Infect Immun* 59, 1991, p. 3261.

FIBRONECTIN-BINDING PROTEINS IN GROUP C STREPTOCOCCI - IDENTIFICATION OF A COMMON STRUCTURAL BINDING DOMAIN

Per-Eric Lindgren[1], Martin McGavin[2,3,] Christer Signäs[1], Bengt Guss[1], Liisa Rantamäki[1], Hans-Peter Müller[1], Magnus Höök[2,4] and Martin Lindberg[1,5]

ABSTRACT

Binding of bacteria to fibronectin (Fn) has been implicated as a mechanism of bacterial adhesion to the host tissue. We have isolated and characterized the genes encoding Fn- binding proteins from two streptococcal species of Lancefield serological group C, *Streptococcus dysgalactiae* (two genes) and *Streptococcus equisimilis* (one gene). The three Fn-binding proteins, which have deduced molecular masses 117-123 kDa, show characteristics of Gram-positive cell surface proteins and the amino acid composition is similar. In all three proteins the binding site to the 29 kDa NH_2-terminal Fn fragment has been localized to repeated domains in the COOH terminal parts. Each repeat is 32-43 amino acids long and occurs 3-4 times consecutively. The repeats show a high degree of similarity between the different species indicating their coinciding function. Synthetic peptides mimicking each of the repeated regions from *S. dysgalactiae* and modifications of them have been made. Conserved, short stretches of amino acids within the repeats, necessary for the Fn

1 Swedish University of Agricultural Sciences, Department of Microbiology, Box 7025, S-750 07 Uppsala, Sweden

2 University of Alabama at Birmingham, Department of Biochemistry, 1918 University Blvd., BHSB 508, Birmingham, Alabama 35294 USA

3 Department of Medical Microbiology, University of Manitoba, 730 William Ave., Winnipeg, Manitoba, R3E 0W3, Canada (present address)

4 Institute of Biosciences and Technology, Texas A&M University, Houston, TX 77030 USA

5 Corresponding author

binding, have been identified. Similar conserved regions are also present in the Fn-binding proteins from *Streptococcus pyogenes* and Staphylococcus aureus.

1. INTRODUCTION

Fibronectin (Fn) is a glycoprotein with a molecular mass of 450 kDa, consisting of two subunits linked by disulfide bonds near the COOH-terminal end of the molecule (1). The protein is distributed throughout the animal body, and is found as a soluble protein in high concentrations in body fluids such as blood and in an immobilized form in the extracellular matrix of various tissues. Adhesion of eukaryotic cells to Fn primarily involves a family of cell surface receptors called integrins, which recognize and bind to specific sites in the central part of the protein. A primary recognition site for the integrin receptors is defined by the amino acid sequence RGD, and synthetic peptides based on this motif inhibit the attachment of many cell types to Fn substrates (2).

The binding of bacteria to Fn was first reported by Kuusela (3), who described its interaction with *Staphylococcus aureus.* A staphylococcal Fn-binding protein was isolated (4, 5), and a gene encoding a protein with Fn binding activity was cloned (6) and sequenced (7). Later a second gene from the same strain encoding a Fn-binding protein was characterized (8). Other Gram-positive bacteria such as group A, C and G streptococci (9, 10) and coagulase negative staphylococci (11) have been shown to bind Fn. Also several species of gram-negative bacteria, e. g. *Escherichia coli* (12, 13, 14), *Salmonella dublin* (15) and *Treponema pallidum* (16) have been reported to bind Fn. Many of these bacterial species which bind Fn are common wound pathogens and it has been suggested that Fn serves as a substratum for bacterial adherence in the damaged tissue. Binding by the bacteria of Fn and also binding of other extracellular matrix proteins, may be the first step in the development of an

2. MATERIALS AND METHODS

2.1. Bacterial strains, plasmids and culture media

Animal derived isolates of *Streptococcus dysgalactiae* and *Streptococcus equisimilis*, Lancefield group C streptococci, were obtained from the Department of Veterinary Microbiology, Swedish University of Agricultural Sciences, Uppsala, Sweden and from the Department of Microbiology and Epizootology, College of Veterinary Medicine, Helsinki, Finland. *E. coli* K12 strain TG1 (20) was used as a host for transformations with the plasmid vector pUC18 (21). The streptococcal and staphylococcal strains were grown in Todd-Hewitt broth (Difco laboratories, Detroit, MI, USA) and *E. coli* was grown in Luria broth. All cultures were grown at 37°C. Bacteria were stored at -70°C in their respective media supplemented with 20% glycerol.

2.2. Restriction endonucleases and other enzymes

Restriction enzymes and T4 DNA-ligase were purchased from International Biotechnologies Inc. (New Haven, CT, USA), New England Biolabs (Beverly, MA, USA), Promega (Maidson, WI, USA) or Boehringer Biochemicals (Mannheim, FRG) and used according to the suppliers' recommendations. For exonuclease-III digestions an Erase-a-Base system kit from Promega was used following a standard protocol from the supplier. Other methods involving DNA techniques were essentially as infection (17, 18).

With the aim to further investigate the mechanisms of the Fn binding in group C streptococci we have characterized three different Fn-binding proteins from the animal pathogens *S. dysgalactiae* and *S. equisimilis* and compared their structural organization to the Fn-binding proteins from *S. pyogenes* (19) and *Staphylococcus aureus* (7, 8).

described (22). For polymerase-chain-reaction the Gene Amp™ kit, obtained from Perkin Elmer Cetus, was used.

2.3. Purification and radiolabelling of fibronectin and fibronectin fragments

Purification of human fibronectin (Fn) from plasma was based on affinity chromatography on gelatin-Sepharose (23). The 29 kDa NH_2-terminal Fn fragment was generated by thermolysin digestion of Fn and purified as described in Lindgren et al (24). The proteins were labeled with ^{125}Iodine (Amersham Inc., Amersham, England), using the Chloramine-T method (25).

2.4. Assays for fibronectin binding

2.4.1. Direct binding assay

S. dysgalactiae S2 was grown over night in Todd-Hewitt broth. Bacteria were collected by centrifugation, and suspended in phosphate buffered saline (PBS: 140 mM NaCl, 10 mM phosphate, pH 7.4) to a density of 1×10^{10} cells x ml^{-1}. Bacteria (5×10^8 CFU/ml) were incubated with ^{125}I-labeled Fn in 0.5 ml of PBS containing 0.1% bovine serum albumin (BSA) and 0.1% Tween 80 to block unspecific binding to cells and tubes. The mixture was rotated in an end-over-end mixer for 1 h at 22°C. Incubation was stopped by the addition of 2.5 ml of ice-cold 0.1% Tween 80 in PBS, and the tubes were centrifuged at 1,400 x g for 15 min. After removal of the supernatant, the radioactivity associated with the cells was quantified in a gamma counter.

2.4.2. Inhibition assay

Solubilized proteins were analysed for Fn-binding activity by measuring their ability to compete with streptococcal cells of the respective species for binding of ^{125}I-labeled Fn. The assay was performed in the same way as described above for the direct binding, with the exception that aliquots of soluble protein were added to the assay mixture.

2.5. Preparation of DNA and DNA sequencing

For preparation of chromosomal DNA, cells of the streptococcal species were grown overnight in Todd-Hewitt medium supplemented with 0.6% yeast extract and 10 mM glycin. The cells were washed in 0.05 x TE(1 x TE: 1.0 M Tris-HCl pH 7.0 + 1.0 M EDTA)-buffer and suspended in 0.05 x TE containing 25% sucrose. Lysozyme was added to a final concentration of 30 mg/ml and the suspensions were incubated for 2 hours at 37°C. The cells, now mainly protoplasts, were washed in 0.05 x TE + 25% sucrose and incubated in 1% SDS + 0.05 x TE for 15 minutes at 65°C and the cell debris were removed by centrifugation. The viscous supernatants were then treated as described for chromosomal DNA preparations (22). Plasmid DNA from *E. coli* clones was prepared, using a boiling method including LiCl extraction (26) or using a cleared lysate method (22).

Nucleotide sequences were determined by the dideoxy chain-termination method of Sanger *et al.* (27). The sequence reactions were performed using "Sequenase, version 2.0" kit (United States Biochemical Corporation, Cleveland, Ohio, USA). The DNA sequencing samples were analyzed by PAGE on wedge-shaped 6% acrylamide gels containing 7.0 M urea (28). The sequence data were recorded and analyzed by different computer programs (29).

2.6. Hybridization methods

Colony hybridization experiments were performed as described by Sambrook *et al.* (22). Plasmid DNA was denatured in 0.2 M NaOH, diluted and bound to a nitrocellulose filter using a dot blot apparatus (Bio-Rad, Richmond, CA, USA). The restriction fragments used as probes were derived from clones containing genes *fnb*A and *fnb*B from *S. dysgalactiae* strain S2, gel purified and labelled with ^{32}P-ATP by a random-priming method (Multiprime DNA labelling system, Amersham Inc, Amersham, England). After hybridization at 42°C overnight the filters were washed with 0.1 X SSC supplemented with 0.1% SDS at 55°C. Southern blot hybridization was performed essentially as described (22, 30).

3. RESULTS

3.1. Binding of fibronectin to *Streptococcus dysgalactiae* strain S2

In agreement with previous studies (31) strain S2 expressed Fn-binding protein. A binding site for *S. dysgalactiae* was localized to the NH_2-terminal domain of Fn since a ^{125}I-labeled fragment encompassing this region, like intact Fn, bound to the bacterial cells in a time dependent process. Furthermore, the bacteria could be saturated with labeled ligand and digestion of cells with trypsin resulted in reduction of the ability to bind Fn. The bacteria were also incubated with ^{125}I-labeled intact Fn or the 29 kDa NH_2-terminal tryptic fragment of Fn in the presence of increasing amounts of the corresponding cold protein or peptide. The cold 29 kDa fragment inhibited the binding of both ^{125}I-labeled molecules to the same degree as cold intact Fn.

A lysate of *S. dysgalactiae* S2 prepared by sonicating the bacterial cells in a buffer containing protease inhibitors inhibited the binding of ^{125}I-labelled intact Fn or the 29 kDa Fn fragment in a concentration dependent manner. When the lysate was passed over a column of Fn-Sepharose, the inhibitory activity was removed. On the other hand, when the lysate was passed through a column of albumin-Sepharose, the activity was only marginally reduced. These results suggested that the inhibitory activity of the lysate was due to the presence of Fn-binding proteins.

3.2. Construction of a genomic library from *Streptococcus dysgalactiae* strain S2 in *E. coli* TGI

Chromosomal DNA from *S. dysgalactiae* S2 was partially digested with *Sau3*AI and fractionated on an 1% agarose gel. Fractions, containing DNA of 3 to 9 kilobase pairs (kb) were mixed with pUC18 vector. The mixture was ligated with T4 DNA ligase, and used for transformation of *E. coli* TG1 selecting for ampicillin resistance. Seven hundred white colonies were collected using blue/white screening, replicaplated to nitrocellulose membranes and after growth the bacterial colonies were lysed. After fixation of DNA the membranes were screened with ^{125}I-labelled 29 kDa Fn fragment.

Three positive clones, with recombinant plasmids called pSDF100, pSDF200 and pSDF300, were identified. Since the restriction cleavage patterns of the inserts in two of the recombinant plasmids, pSDF100 and pSDF300, were partially overlapping, only pSDF100 (with an insert of 4.9 kb), encoding the highest Fn-binding activity of the two, and pSDF200 (with an insert of 6.9 kb) were further studied. Figures 1A and 1D show the completely different restriction maps of the inserts in pSDF100 and pSDF200.

3.4. Subcloning and sequencing of genes

The genes of the original clones, containing plasmids called pSDF100 and pSDF200 (24), were further subcloned using appropriate restriction endonuclease cleavages and digestion with exonuclease III for different time periods. In this way a series of overlapping subclones were constructed (Fig. 1C and Fig. 1F). Both strands of the inserts in the plasmids were sequenced. The whole gene, *fnb*A, coding for the Fn-binding A-protein (FnBA), was present within the insert of streptococcal DNA in pSDF100 (Fig. 1C). The second gene, *fnb*B, coding for the Fn-binding B-protein (FnBB), was incomplete in the original insert in pSDF200. Thus, in order to obtain the sequence of the very last 3'-end of the gene (Fig. 1F), it was necessary to screen a library prepared by cleaving the chromosomal DNA with the restriction enzymes *Xho*I and *Hind*III, and the pUC18 vector with *Hind*III and *Sal*I. The library was screened with a ^{32}P-ATP labelled 20mer synthetic oligonucleotide of a unique sequence derived from a site near the 3'-end of the insert in pSDF201. One positive clone containing a plasmid called pSDF250 was chosen for further studies (Fig. 1F).

3.5. Nucleotide sequences of genes *fnb*A and *fnb*B and the deduced amino acid sequences of encoded proteins

The DNA sequence of the insert in pSDF100 containing *fnb*A has an open reading frame, starting with an ATG codon at nucleotide 437 and terminating in a TAA stop codon at nucleotide 3710, encoding a polypeptide of 1091 amino acids.[1] Downstream of the coding region, there is a hairpin loop structure that may act as a transcriptional terminator. As the encoded protein is a cell-surface protein in *S. dysgalactiae*, a signal peptide should precede the mature protein, which seems to be the case. There is a possible cleavage site for signal peptidase after amino acid residue numer 24. One major region of repetitive units was identified by computer analysis comparing the sequence of the gene with itself in a dot plot matrix. The repetitive region, called R_A, is located near the 3'-end of the gene (Fig. 1B)and consists of a motif of 32-43 amino acids repeated three times (Fig. 2). Furthermore, starting twentyone amino acids upstream of R_{A1} there is a region of twelve amino acids very similar to the COOH-terminal parts of the R_A-repeats. Flanking the R_A-region there are domains rich in prolines. The upstream region is a unique stretch of approximately 50 amino acids called X_{AP} of which 16 amino acids are prolines. Downstream of the R_A-region, the proline residues are again more frequent than in the rest of the protein, which is typical for the cell-wall-spanning domain (W_A) of studied cell surface proteins in Gram-positive cocci (32, 33).

The insert in pSDF201 contains an open reading frame encoding the main part of *fnb*B while the insert in pSDF250, which partly overlaps pSDF201, contains an open reading frame encoding the very last 3'-end of *fnb*B (Fig. 1F). The combined sequences of the inserts in pSDF201 and pSDF250 give the nucleotide sequence of the complete gene *fnb*B.[2] The insert contains an open reading frame, coding for a polypeptide of 1117 amino acids, beginning with a TTG start codon at nucleotide 140 and terminating in a TAA stop codon at nucleotide 3491 (Fig.3). There is one possible cleavage site for a signal peptidase resulting in a signal peptide (S_B) of 27 amino acids. Downstream of the coding region, there is a loop structure, which probably acts as a transcription terminator. As in gene *fnb*A, a repetitive domain, called R_B, is present in the 3'-end of the gene. This part of the gene encodes a region containing a unit of 36 amino acids repeated twice and partly a third time (Fig. 2). Downstream of the repeat sequence there is a proline-rich region, W_B, probably spanning the cell wall.

The COOH-terminal ends of the FnBA and FnBB proteins have similar general features, including the consensus sequence LPXTGX invovled in anchoring of the protein to the peptidoglucan as suggested by Fischetti *et al.* (34), and hydrophobic transmembrane domains (M_A and M_B) followed by a short stretch of charged, mainly basic, amino acids. All these features are general characteristics of cell surface proteins of Gram-positive cocci.

3.6. The fibronectin-binding regions of FnBA and FnBB from *S. dysgalactiae*

The inserts of the smallest subclones, obtained using exonuclease III digestions, called pSDF102c14 and pSDF203c8, encoding Fn-binding peptides, consisted each of approximately 500 bp. Both clones contained at least parts of the repetitive regions, which were lacking in the negative subclones, thus indicating that the ability to bind Fn was localized to the R_A and R_B regions. In order to confirm this, the entire R_A- and R_B-regions (Figs 1B and 1E) were separately cloned using PCR-technology. The R_A- and R_B- regions were amplified using two different synthetic oligonucleotide primers, respectively (35).

The amplified R_A-and R_B-regions were analysed using agarose-gel electrophoresis. The amplified sequences were cleaved with *Bam*HI and *Hind*III, and the respective fragments were ligated into pUC18 (previously digested with the same enzymes). After ligation and transformation into *E.coli* TG1 cells, two clones, called pR_A100 and pR_B200, were further studied. The inserted fragments were sequenced to confirm that they were correctly amplified. The pR_A100 and pR_B100 were digested with

*Bam*HI and *Hind*III and the inserted fragments purified and ligated into the gene fusion/secretion vector pEZZ18 (36), previously cleaved with the same enzymes. After transformation into *E.coli* TG1 cells two clones called $pZZR_A100$ and $pZZR_B200$ were tested for their ability to encode fusion proteins with Fn-binding properties. Since the two different constructs should encode extracellular fusion proteins with the ability to bind to IgG, cellfree growth media from overnight cultures were passed over IgG-Sepharose columns to affinity purify the fusion proteins (36). The bound proteins were eluted with 1.0 M HAc, pH 2.8, lyophilized and analysed using SDS-PAGE (Fig. 3). The samples were loaded in duplicates on the gel and after electrophoresis the gel was divided into two halves. One was stained with Coomassie Brilliant Blue (Figure 3A) and the other was electroblotted to a nitrocellulose filter (Western blotting) and probed with ^{125}I-labelled NH_2-terminal fragment of fibronectin (Fig. 3B). The affinity purified fusion proteins, encoded by $pZZR_A100$ and $pZZR_B200$, with deduced molecular masses of 28 and 27 kDa respectively, were homogenous in size on the SDS-gel but the molecular masses were considerably higher than the deduced ones (Fig 3A, lanes 3 and 5). The reason for this discrepancy is not clear but similar results have been reported for other bacterial cell wall proteins (32, 7). In Western blots the fusion proteins bound the ^{125}I-labeled 29kDa fragment of Fn (Fig. 3B, lanes 1 and 3). The respective purified fusion proteins also inhibited the binding of the same fragment of Fn to *S.dysgalactiae* strain S2 and to *S.aureus* strain Cowan I cells, when tested in an inhibition assay. *E.coli* cell lysates from clones containing the plasmids pSDF102c9 and pSDF203, which were used as templates in the PCR amplifications, also contained peptides with Fn-binding activity (Figs 5A and 5B, lanes 2 and 4).

3.7. Inhibitory activity of synthetic peptides

Peptides, mimicking each of the type R_A structural motifs in protein FnBA and the type R_B motifs in protein FnBB, were synthesized (37) and assayed for Fn-binding activity by measuring their ability to inhibit binding of ^{125}I-labelled intact Fn or 29 kDa Fn fragment to cells of *S. dysgalactiae* as described by McGavin *et al.*, (38). Peptides A1, A2 and A3, which correspond to the different units of the R_A-region, all inhibited the binding of ^{125}I-labelled intact Fn (data not shown) or the 29 kDa NH_2-terminal Fn fragment (Fig. 4A) to cells of *S. dysgalactiae* in a concentration dependent manner. The A peptides were more efficient inhibitors when the isolated NH_2 terminal Fn fragment was used as a ligand compared to intact Fn, and A2 was consistently more effective than A1 and A3. Fifty percent inhibition of binding of the 29 kDa fragment to bacteria was observed with 200 μM of A1, 2 μM of A2 and 80 μM of A3.

When the B peptides, which correspond to the different units of the R_B region, were tested for inhibitory activity, only peptide B3 inhibited the binding of ^{125}I-labelled intact Fn or its NH_2-terminal fragment to cells of *S. dysgalactiae* (Fig. 4B). Neither B1 or B2 showed any activity at the concentrations tested. Again, this was more apparent when the 29 kDa fragment was used in the inhibition assays instead of intact Fn. Forty μM of B3 caused a fifty percent inhibition of binding of the NH_2-terminal fragment to the bacterial cells, which suggested that the potency of B3 is between that of A2 and A3.

3.8. Fibronectin-binding peptides have acidic residues essential for activity

In a previous study with the D3 structural motif of the Fn-binding A-protein (FnBPA) from *S. aureus*, we proposed that the sequence DFEEDT within D3 was a major determinant of Fn binding (37). This was based on observations that chemical modification of carboxyl side chains, or proteolytic cleavage within this sequence resulted in loss of binding activity. Furthermore, a synthetic peptide, S16-36, representing amino acids 16-36 of the 37 amino acids long D3 motif, retained binding activity which was only two or three fold less than the intact peptide. The only acidic residues in the truncated peptide were contributed by the DFEEDT sequence. Although the amino acid sequences of the A and B motifs of the *S. dysgalactiae* FnBA and FnBB proteins were very different , they both contained acidic core sequences, DFTEDT/S and EVEDT/S, respectively, resembling that of the D3 motif of *S. aureus* FnBPA.

Chemical modification through 1-ethyl-3-(3-dimethylaminopropyl carbodiimide) (EDC) mediated condensation of glycine methyl ester on carboxyl groups of aspartate and glutamate residues resulted in loss of activity for both peptides, whereas dihydroxypropylation-reduction of lysine caused only marginal loss of activity. Hence we concluded that one or more of the acidic residues in each of the A2 and B3 motifs is essential for ligand binding. (38)

Within motif A2, there are eleven acidic residues. When A2 was cleaved with cyanogen bromide (Fig 5a), one active fragment, CB1-30 was recovered (Fig. 5b panel A). This fragment contains eight acidic residues, and elicited a 50% inhibition of binding of the ^{125}I-labelled 29 kDa Fn fragment to S. dysgalactiae cells at a concentration of 130 ug/ml. With the control peptide A2, 50% inhibition was achieved at a concentration of less than 20 μg/ml. Since CB1-30 showed significant activity, the acidic residues essential for activity appeared to be contained within this domain. When peptide A2 was cleaved with chymotrypsin, the products CT1-22, and CT23-43 were inactive at concentrations as high as 200 μg/ml (Fig. 5b panel A). Since CT cleaves within the DFTDET sequence we suggest that the integrity of this sequence is required for activity, and that one or more essential residues are present.

Fragmentation studies were also conducted with peptide B3 (Fig. 5a). Chymotrypsin cleavage at Phe_{28} yielded an active fragment CT1-28, and an inactive fragment, CT29-36 (Fig. 5b panel B). The amount of peptide required to elicit a 50% inhibition of ^{125}I-labelled 29 kDa Fn fragment binding to cells of *S. dysgalactiae* was 170 μg/ml and 80 μg/ml for CT1-28 and B3, respectively. Trypsin cleavage products T1-23, and T24-36 were both inactive at concentrations up to 200 μg/ml (Fig. 5b panel C). In the active fragment, CT1-28, there are six acidic residues. However all of these are also found within the inactive cleavage product T1-23. Therefore residues other than the essential acidic amino acids are also required for Fn binding.

3.9. Some conserved residues important for activity

To further examine the importance of different residues for binding in CB1-30 of the A2 motif (Fig. 5a), we synthesized a series of "mutants" in which various discrete residues have been substituted. After purification, these peptides were tried as inhibitors of the binding of biotin labelled recombinant S. aureus FnBP to human Fn immobilized in microtitre wells as described by McGavin *et al.* (38) . The results of these experiments showed that at 1 mg/ml of CB1-30 in A2 inhibits the binding of this receptor to Fn by 67%. Substitution of residues ED, S or EDE in positions 1-2, 3 and 8-9-10, respectively, did not significantly alter the inhibitory activity of the peptide. However, when one or the other or both of the GG in position 14-15 were replaced, the resulting peptides were essentially inactive. Furthermore, substituting the hydrophobic residues at positions 20 and 22 with a hydrophilic residue resulted in substantial loss of activity, whereas replacing F_{22} with I_{22} was accompanied with only a minor activity reduction. The acidic residues essential for activity are apparently located in the COOH-terminal half of the molecule and replacing E_{24}, D_{25} resulted in loss of activity whereas $D_{21} \rightarrow N_{21}$ only reduced the activity slightly. Introducing an additional acidic residue at position 23 did not significantly affect the activity.

3.10. Cloning of a gene encoding a fibronectin-binding protein from *Streptococcus equisimilis* strain Se165

DNA fragments covering the repeat regions of *fnb*A and *fnb*B coding for the Fn-binding activity of *S. dysgalactiae* strain S2 (35), were used as probes in hybridization experiments for detection of similar genes in strains of *S. dysgalactiae*, *S. equi*, *S. equisimilis* and *S. zooepidemicus*. Both gene probes hybridized to 12 tested strains of *S. dysgalactiae*. The probe derived from *fnb*B, but not the one from *fnb*A hybridized to all 12 tested strains of *S. equisimilis*. The four strains of *S. equi* and two strains of *S. zooepidemicus*, were negative in the hybridization tests indicating that these species possess Fn-binding proteins which are not closely related to the binding proteins of *S. dysgalactiae*.

A genomic library from *S. equisimilis* strain Se165 was constructed by cleavage of chromosomal DNA with *Nhe*I, based on previous tests of cleavage patterns with different restriction endonucleases and hybridizations using an *Acc*I-*Acc*I fragment from *S. dysgalactiae* as probe. The cleaved DNA was fractionated on an 1% agarose gel and fragments of 3.5 to 4.5 kilobasepairs (kb) were electroeluted and mixed with *Xba*I-cleaved pUC18 DNA, pretreated with alkaline phosphatase. The mixture was ligated with T4 DNA ligase and transformed to *E. coli* K12 strain TG1 with selection for ampicillin resistance. Colonies containing streptococcal DNA inserts were detected using blue/white screening, collected and probed by using the *Acc*I-*Acc*I fragment. Several positive clones were obtained and the isolated plasmids all showed the same restriction pattern. The clone, containing the plasmid pSEF100, with a 4.0 kb streptococcal insert was chosen for further studies. A lysate from the clone inhibited the binding of 29 kDa Fn-fragment to *S. equisimilis* indicating the expression of a Fn-binding protein. Probes selected from pSEF100 hybridized to chromosomal DNA from *S. equisimilis* Se165, verifying that the cloned DNA was derived from this strain. The smallest positive subclone, called pSEF400, contained a 1.2 kb *Hind*III-*Hind*III fragment from pSEF100.

3.10. Western blotting of Fn-binding protein from *S. equisimilis*

A Western blot analysis for detection of Fn-binding proteins in crude cell extracts and ammonium sulfate precipitates of the culture supernatants from *S. equisimilis* strain Se 165 and the clones containing pSEF100 and pSEF400 is shown in Figure 6. In the crude cell extract of Se165 a major Fn-binding component was present with a molecular mass of approximately 140 kDa which is in the same range as the deduced molecular mass. A polypeptide of similar molecular mass was also clearly visible in lane E with precipitated culture supernatant from the same strain. In cell extracts from the clone containing pSEF100 (lane B) a polypeptide of similar size was detected, although in the crude culture supernatant also several smaller Fn-binding components were present. From the clone containing pSEF400 (lane C) at least three separate Fn-binding peptides, in the range of 40-60 kDa, were obtained.

3.11. Nucleotide sequence of gene *fnb* from *S. equisimilis* and the deduced amino acid sequence of the encoded protein

The nucleotide sequence of both strands of a 3584 bp long part of the insert in pSEF100 was determined.[3] The sequenced part of the insert in pSEF100 contains an open reading frame of a gene encoding a polypeptide of 1092 amino acids. As the encoded protein is a cell-surface protein in *S. equisimilis,* a signal peptide should precede the mature protein. One possible cleavage site for a signal peptidase is indicated resulting in a signal peptide of 27 amino acids and a structural protein with a molecular mass of 120 kDa. In the COOH-terminal part of the deduced amino acid sequence there is a domain containing the 36 amino acids long repeat units, R_{E1}-R_{E3}, followed by a short, proline-rich cell wall spanning domain, W_E, and a domain with mostly hydrophobic amino acids spanning the cell membrane, M_E, followed by a short tail of seven, mainly positively charged amino acids in the COOH-terminal end of the protein. The M_E-region is preceded by the hexapeptide LPATGE, corresponding to the concensus sequence LPXTGX considered to be involved in the anchoring of cell surface proteins to the cell wall in Gram-positive cocci (34).

3.12. Comparison of the deduced amino acid sequences from FnB of *S. equisimilis* and FnBB from *S. dysgalactiae*

The deduced amino acid sequence encoded by the gene *fnb* from *S. equisimilis* Se165 was compared to the corresponding sequence encoded by *fnb*B from *S. dysgalactiae* S2 (Fig. 7). The signal peptides are very similar, only 4 of 27 amino acids are different. In the NH_2-terminal parts of the structural proteins (X-regions) the similarity is generally low but there are some short homologous stretches of 4-5 amino acids. Downstream from the central parts of the proteins to the COOH-terminal ends, including the Fn-binding domains, the similarity is again around 90%.

4. CONCLUSION

Two genes from *S. dysgalactiae* and one from *S. equisimilis* encoding Fn-binding proteins have been cloned and sequenced (24, 35, 39). All three proteins have structural features typical of cell-wall associated proteins in Gram-positive cocci (40). Sequences typical of signal peptides are present in the NH_2-terminal end and they have similar hydrophobic membrane spanning domains and a number of basic amino acid residues in the COOH-terminal end. Upstream of the membrane spanning domain there is a hexapeptide sequence LPXTGX, probably involved in cell-wall anchoring of the proteins (34). However, in one of the *S. dysgalactiae* proteins (FnBB) there is an alanine (A) instead of a threonine (T) at position four, which is also the case in protein L, an immunoglobulin binding protein from *Peptostreptococcus magnus* (41). A wall-spanning domain, rich in prolines, is recognized, and directly upstream this domain the Fn-binding domains, consisting of 2.5-4 repeats of 32-43 amino acids, are located. Within these repeats two glycine residues are present next to each other, surrounded by some acidic amino acid sequences (Fig. 8). The motives EDT, EED and TED are found several times in these repeats from all species. These acidic stretches may play a central role for formation of the three-dimensional structure and thus be of great importance for the ability to bind Fn. Direct upstream the Fn-binding domains there is no unique sequence except of a 50 residues long region, X_{AP}, in one of *S. dysgalactiae* proteins (FnBA), in which every third amino acid is a proline. A similar sequence has been observed in an IgA-binding protein from group B streptococci, called Bac, in which there is a 90 residue segment where every third residue is a proline (42).

A comparison between the Fn-binding repeats of *S. dysgalactiae* FnBA and FnBB (24), and *S. pyogenes* Sfb (19), including flanking regions, is shown in Figure 8. Apparent sequence similarities are present within the binding domains and the regions downstream of these. It is worthwhile to note that the sequence similarity between Sfb from *S. pyogenes* and the corresponding *S. dysgalactiae* proteins is higher than between the *S. dysgalactiae* proteins themselves.

Fn-binding activity is encoded by more than one gene in several Gram-positive cocci. Since these proteins have been proposed and also shown to be of importance for the adherence and virulence of the bacteria (17, 18, 43, 44), increased knowledge about their structure and function will facilitate applications of these proteins in the prevention of bacterial infections.

5. ACKNOWLEDGEMENTS

This investigation was supported by grants from the Swedish Medical Research Council (B91-16X-03778-19 and B92-16X-09923-01A), the Swedish Council for Forestry and Agricultural Research (32.0357/92), the National Institute of Health (AI20624), and Alfa Laval AB, Tumba, Sweden. Martin McGavin was supported by a post doc fellowship from the Medical Research Council of Canada.

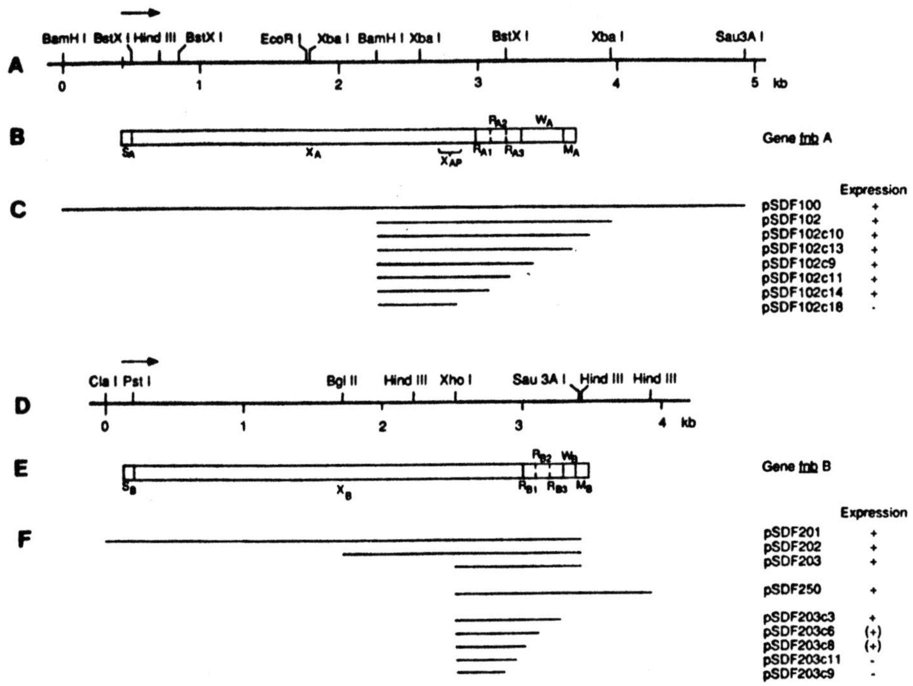

Figure 1. Restriction maps of inserts and schematic diagrams of proteins, clones and subclones. (A) Restriction map for gene *fnb*A. The arrow indicates the direction of transcription of the gene. (B) Schematic diagram of the protein encoded by gene *fnb*A. S, signal sequence; X, spacer region; X_{AP}, proline-rich sequence of the spacer region; R, domain consisting of repeated motifs; W, cell-wall-spanning domain; M, membrane-spanning domain. (C) Different clones and subclones. The Fn-binding activities of lysates from the different subclones, expressed as inhibition of binding of rediolabelled Fn to *S. dysgalactiae* S2, are indicated. More than 85% inhibition, +; less than 15%, -. (D) Restriction map for *fnb*B. (E) Schematic diagram of the protein encoded by gene *fnb*B. (F) Clones and subcloned for *fnb*B. Explanation as in A-C for *fnb*A but (+) indicates 30-40% inhibition. (35)

FnBA

A1	EDTQT SQEDI V-LGG PGQVI DFTED SQPGM SGNNS HTIT
A2	EDSKP SQEDE VIIGG QGQVI DFTED TQSGM SGDNS HTGDT VLE
A3	EDSKP SQEDE VIIGG QGQVI DFTED TQTGM SGAGQ VESP

FnBB

B1	EETLP TEQGQ SGSTT EVEDT KGPEV IIGGQ GEIVD I
B2	EENLP TEQGQ SGSTT EVEDT KGPEV IIGGQ GEVVD I
B3	EESLP TEQGQ SGSTT EVEDS KPKLS IHFDN EWPKE D

Figure 2. The repetitive domains of FnBA and FnBB from *S. dysgalactiae*. The identical amino acids are shadowed. The conserved octapeptide EVIIGGQG is underlined. (35, modified from figure 4.)

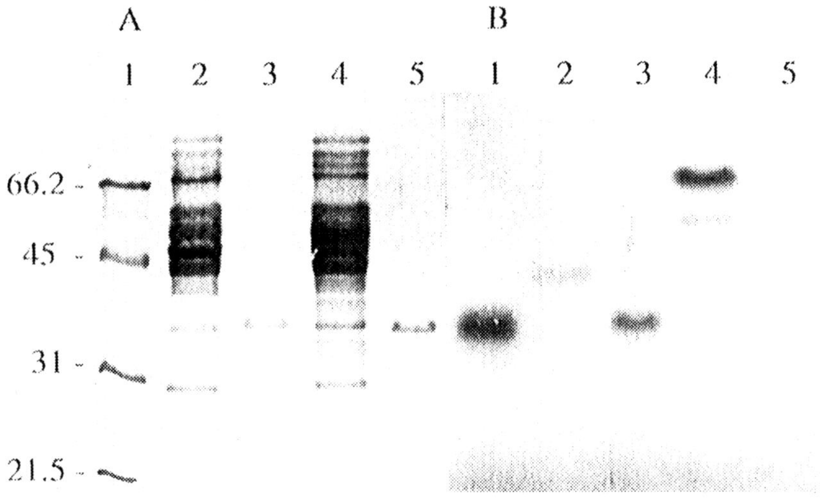

Figure 3. Identification of Fn-binding activity in crude cell lysates and culture media of different *E. coli* clones. SDS-PAGE was performed using stacking and running gels with 4% and 12% acrylamide respectively (45). After SDS-PAGE one half of the gel was stained with Coomassie blue (A). The other half of the gel was electrophoretically transferred to a nitrocellulose membrane (Schleicher and Schüll) and probed with ^{125}I-labelled 29 kDa fragment of Fn (B). Proteins used as molecular size markers are indicated in (A), lane 1: bovine serum albumin (66.2 kDa); hen egg-white ovalbumin (45 kDa); bovine carbonic anhydrase (31 kDa); soybean trypsin inhibitor (21.5 kDa). (A) Lane 2 and (B) lane 4, samples from *E. coli* clone containing pSDF102c9; (A) lane 3 and (B) lane 3, samples from *E. coli* clone containing pZZR$_A$100; (A) lane 4 and (B) lane 2, samples from *E. coli* clone containing pSDF203; (A) lane 5 and (B) lane 1, samples from *E. coli* clone containing pZZR$_B$200. (35)

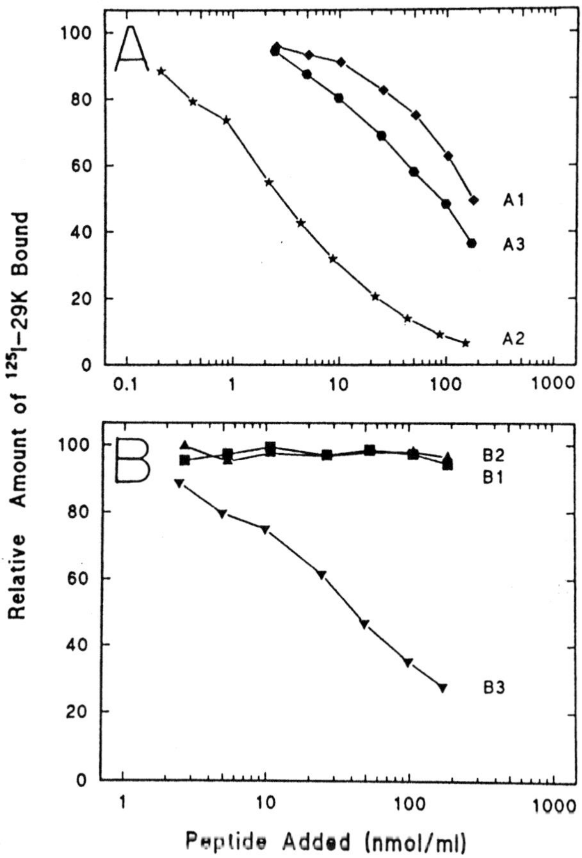

Figure 4. Inhibition of ^{125}I-labelled 29 kDa fragment of Fn to *S. dysgalactiae* S2 by addition of synthetic peptides mimicking the individual repeated domains of FnBA (A) and FnBB (B). The indicated amount of peptide was mixed with 5 x 10^8 cells and 5 x 10^4 cpm of ^{125}I-labelled 29 kDa Fn fragment in an assay volume of 0.5 ml of PBS supplemented with 0.1% bovine serum albumin and 0.1% Tween 80. After 60 minutes of end-over-end incubation at room temperature, unbound Fn was diluted by adding 3 ml ice cold PBS containing 0.1% Tween 80. After centrifugation the radioactivity associated with the bacterial pellet was quantitated using a LKB gamma counter. The values are expressed as percentage of the radioactivity bound in the absence of added peptide. (38)

```
FnBA
     A2   EDSKPSQEDEVIIGGQGQVIDF TEDTQSGM SGDNSHTGDTVLE
                                ▲        ▲
                                CT       CB

FnBB
     B3   EESLPTEQGQSGSTTEVEDSKPK LSIHF DNEWPKED
                                 ▲     ▲
                                 T     CT
```

Figure 5a. Cleavage sites for chymotrypsin (CT), trypsin (T) and cyanogen bromide (CB) are indicated by arrows in the sequences of peptides A2 and B3.

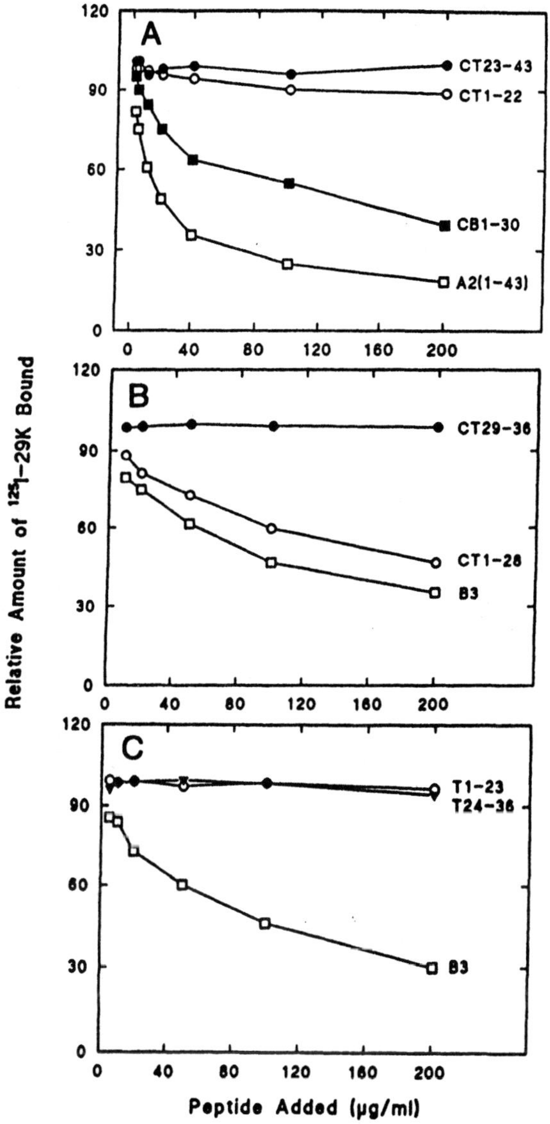

Figure 5b. Inhibitory activity of purified chymotrypsin (CT) and cyanogen bromide (CB) cleavage products of A2 (A), CT cleavage products of B3 (B) and trypsin (T) cleavage products of B3 (C). A2 and B3, unmodified peptides. The numbers indicate the NH_2- and COOH-terminal residues of the purified cleavage products. The ligand was 29 kDa Fn fragment. (38)

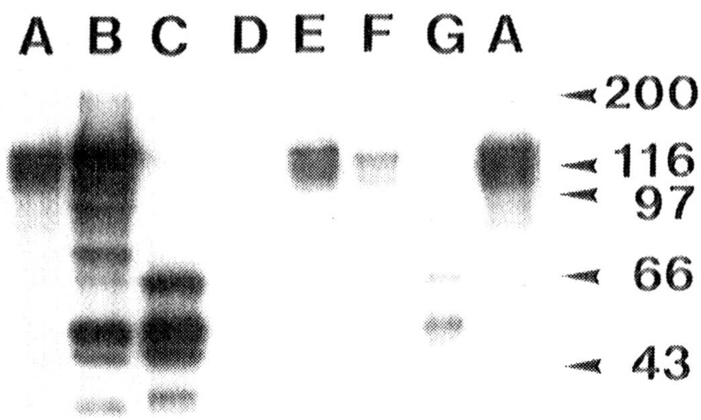

Figure 6. Western blot analysis of Fn-binding proteins released from *S. equisimilis* Se165 and produced by the *E. coli* clones containing pSEF100 and pSEF400. Cells from cultures were centrifugated and the pellets and supernatants were collected separately. The cells were resuspended in PBS, pH 7.4. To all samples 5 mM EDTA and 1 mM PMSF (Phenylmethyl sulfonylfluoride; Merck, Darmstadt, FRG) were added. The cell suspensions were sonicated using a Soniprep 150 sonicator (MSE, Crawley, England) and the cell debris were removed by centrifugation. The proteins in the supernatants were precipitated by adding ammmonium sulfate to 60% saturation and the precipitates were resuspended in water and dialyzed. After boiling in SDS sample buffer with 2-mercaptoethanol and centrifugation the samples were loaded on a 7.5% SDS-PAGE. After electophoresis, the proteins were electrophoretically transferred to nitrocellulose membranes (Schleicher and Schüll, Dassel, FRG), which was saturated 2h with 3% BSA and 0.1% Tween 20 in PBS, and probed with ^{125}I-labelled 29 kDa NH_2-terminal Fn fragment. After washing five times with 0.1% Tween 20 in PBS the membrane was dried and subjected to autoradiography.

Lane A, cell extract of *S. equisimilis* Se165; lane B, cell extract of E. coli clone containing pSEF100; lane C, cell extract of *E. coli* clone containing pSEF400; D, cell extract of E. coli containing pUC18 without insert; E, concentrated culture medium of *S. equisimilis* Se165; F, concentrated culture medium of *E. coli* clone containing pSEF100; G, concentrated culture medium of *E. coli* clone containing pSEF400. Standard molecular mass markers are shown to the right.

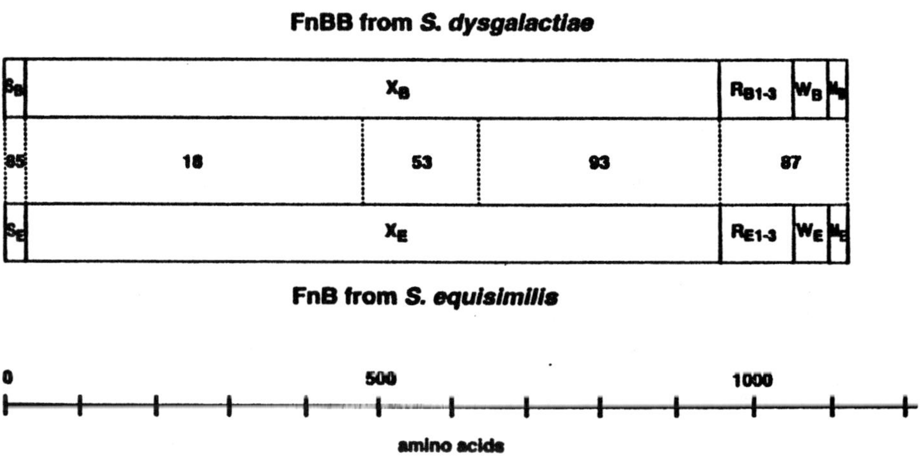

Figure 7. Comparison of the repetitive regions from the Fn-binding proteins of *S. aureus,* FnBPA (7); *S. dysgalactiae,* FnBA and FnBB (35); *S. equisimilis,* FnB (39), and *S. pyogenes,* Sfb (19). The numbers denote the positions in the amino acid sequences of the respective regions. Identical amino acids are indicated by shadowed or bold letters.

```
FnBPA                                                GQNSGNQS FEEDTEED.K
FnBA    VEDTQTSQED IVL.GGPGQV IDFTEDSQPG MSGNNSHTI. .TEDSKPSQE
Sfb     VETEDTK.EP GVLMGGQSES VEFTKDTQTG MSGQTTPQVE .TEDTK...E
FnBB                           EETLPTEQG QSGSTTE.V. ..EDTK...G
FnB                            EETLPTEQG QSGSTTE.V. ..EDTK...V

FnBPA   PKYEQGGNIV DIDFDSVPQ. .IHGQNKGNQ S..FEEDTE. KDKPKYEHGG
FnBA    DEVIIGGQGQ VIDFTEDTQS GMSGDNSHTD GTVLEEDSKP SQEDEVIIGG
Sfb     PGVLMGGQSE SVEFTKDTQT GMSGQ...TA SQVETEDTK. ..EPGVLMGG
FnBB    PEVIIGGQGE IVDIEENLPT EQGQSGSTT. ...EVEDTKG ...PEVIIGG
FnB     PEVMIGGQGE IVEFEETLST EQGQSSSAT. ...EVEDTKG ...PDVLIGG

FnBPA   N..IIDIDFD SVPHIHGFNK HTE.IIEEDT NKDKPSYQFG GHN.SVDFEE
FnBA    QGQVIDFTED TQTGMSGAGQ VESPTITEET HK..PEIIMG GQSDPIDMVE
Sfb     QSESVEFTKD TQTGMS..GQ TTPQVETEDT .KE.PGVLMG GQSESVEFTK
FnBB    QGEVVDIEES LPTEQGQSGS TT...EVEDS .K
FnB     QGEIVEFEET LPTEHGQSGS TT...EVEDS .K

FnBPA   DTLPKVSGQN EGQQTIEEDT
FnBA    DTLPGMSGSN EATV.VEEDT RPKLQ
Sfb     DTQTGMSGFS E.TVTIVEDT RPKLV
```

Figure 8. Schematic drawings comparing the domain organization of FnB from *S. equisimilis* and FnBB from *S. dysgalactiae* (35). The identity (%) of corresponding protein regions is indicated as figures between the two protein bars. S, signal peptide; X, spacer region; R, repeat region; W, cell wall spanning domain; M, membrane spanning domain.

Footnotes

[1] The nucleotide sequence of the gene *fnb*A has been deposited with the EMBL sequence data banks and is available under the accession number Z22150.

[2] The DNA sequence of *fnb*B deposited with the EMBL sequence data banks and is available under the accession number Z22151.

[3] The DNA sequence of the gene *fnb* has been deposited with the EMBL sequence data banks and is available under the accession number **XXX**.

6. REFERENCES

1 R.O. Hynes, 1985. "Molecular biology of fibronectin." *Ann. Rev. Cell. Biol.* Vol. 1, p. 67.

2 E. Ruoslahti, E.G. Hayman, and M.D. Pierschbacher. "Extracellular matrices and cell adhesion." *Arteriosclerosis* Vol. 5, 1985, p. 581.

3 P. Kuusela. "Fibronectin binds to *Staphylococcus aureus*." *Nature* Vol. 276, 1978, p. 718.

4 F. Espersen and I. Clemmensen. "Isolation of a fibronectin-binding protein from *Staphylococcus aureus*." *Infect. Immun.* Vol. 37, 1982, p. 526.

5 G. Fröman, L.M. Switalski, P. Speziale, and M. Höök, "Isolation and characterization of a fibronectin receptor from *Staphylococcus aureus*." *J. Biol. Chem.* Vol. 262, 1987, p. 6564.

6 J.-I. Flock, G. Fröman, K. Jönsson, B. Guss, C. Signäs, B. Nilsson, G. Raucci, M. Höök, T. Wadström, and M. Lindberg. "Cloning and expression of the gene for a fibronectin-binding protein from *Staphylococcus aureus*." *The EMBO J.* Vol. 6, 1987, p. 2351.

7 C. Signäs, G. Raucci, K. Jönsson, P.-E. Lindgren, G.M. Anantharamaiah, M. Höök, and M. Lindberg. Nucleotide sequence of the gene for a fibronectin-

binding protein from *Staphylococcus aureus*: Use of this peptide sequence in the synthesis of biologically active peptides. *Proc. Natl. Acad. Sci. USA* , Vol. 86, 1989, p. 699.

8 K. Jönsson, C. Signäs, H.-P. Müller, and M. Lindberg. "Two different genes encode fibronectin binding proteins in *Staphylococcus aureus*: The complete nucleotide sequence and characterization of the second gene." *Eur. J. Biochem.* Vol. 202, 1991, p. 1041.

9 E.B. Myhre, and P. Kuusela. "Binding of human fibronectin to group A, C and G streptococci." *Infect. Immun.* Vol. 40, 1983, p. 29.

10 W.A. Simpson, and E.H. Beachey. "Adherence of group A streptococci to fibronectin on oral epithelial cells." *Infect. Immun.* Vol. 39, 1983, p. 275.

11 L.M. Switalski, C. Rydén, K. Rubin, Å. Ljungh, M. Höök, and T. Wadström. "Binding of fibronectin to *Staphylococcus* strains." *Infect. Immun.* Vol. 42, 1983, p. 628.

12 G. Fröman, L.M. Switalski, A. Faris, T. Wadström, and M. Höök. "*Escherichia coli* binding to fibronectin: Mechanisms for tissue adherence." *J. Biol. Chem.* Vol. 259, 1984, p. 14899.

13 B. Westerlund, P. Kuusela, T. Vartio, I. van Die,and T.K. Korhonen. "A novel lectin-independent interaction of P fimbriae of *Escherichia coli* with

immobilized fibronectin." *FEBS Lett.* Vol. 243, 1989, p. 199.

14 A. Olsén, A. Jonsson, and S. Normark. "A novel class of surface organelles on *Escherichia coli* that mediates binding to fibronectin." *Nature* Vol. 338, 1989, p. 652.

15 K. Kristiansen, S.B. Baloda, J.L. Larsen, and T. Wadström. "Toxins, putative cell adhesins and fibronectin-binding properties of *Salmonella dublin.*" *Acta Pathol. Microbiol. Scand.* Vol. 95, 1987, p. 57.

16 K.M. Peterson, J.B. Baseman, and J.F. Alderete. "*Treponema pallidum* receptor binding proteins interact with fibronectin." *J. Exp. Med.* Vol. 157, 1983, p. 1958.

17 M. Höök, L. Switalski, T. Wadström, and M. Lindberg. (1989) "Interactions of pathogenic microorganisms with fibronectin." In: Mosher, D. M. (ed.), Fibronectin, Academic Press Inc., San Diego, New York, Berkely, Boston, London, Sydney, Tokyo, Toronto. pp. 295-308

18 E. Hanski, P.A. Horwitz, and M.G. Caparon. "Expression of protein F, the fibronectin-binding protein of *Streptococcus pyogenes* JRS4, in heterologous streptococcal and enterococcal strains, promotes their adherence to respiratory epithelial cells." *Infect. Immun.* Vol. 60, 1992, p. 5119.

19, S. Talay, P. Valentin-Weigand, P.G. Jerlström, K.N. Timmis, and G.

Chhatwal. "Fibronectin-binding protein of *Streptococus pyogenes*: Sequence of the binding domain involved in adherence of streptococci to epithelial cells." *Infect. Immunol.* Vol. 60, 1992, p. 3837.

20 P. Carter, H. Bedonelle, and G. Winter. "Improved oligonucleotide site-directed mutagenesis using M13 vectors". *Nucleic Acids Res.* Vol. 13, 1985, p. 4431.

21 J. Norrander, T. Kempe, and J. Messing, "Construction of improved M13 vectors using oligodeoxynucleotide-directed mutagenesis." *Gene* Vol. 26, 1983, p. 101.

22 J. Sambrook, E.F. Fritsch, and T. Maniatis, 1989. Molecular Cloning. A laboratory manual, second edition. Cold Spring Harbor Laboratory Press, New York.

23 M. Vuento, and A. Vaheri. "Purification of fibronectin from human plasma by affinity chromatography under non-denaturing conditions." *Biochem. J.* Vol. 183, 1979, p. 331.

24 P.-E. Lindgren, P. Speziale, M. McGavin, H.-J. Monstein, M. Höök, L. Visai, T. Kostiainen, S. Bozzini, and M. Lindberg. "Cloning and expression of two different genes from *Streptococcus dysgalactiae* encoding fibronectin receptors." *J. Biol. Chem.* Vol. 267, 1992, p. 1924.

25 W.M. Hunter. 1978. In: Weir, D.M. (editor) Handbook of experimental immunology. Blackwell Scientific, Oxford, pp. 14.1-14.40.

26 H.-J. Monstein, and T. Geijer. "A rapid and inexpensive method for preparing *E. coli* plasmid-DNA." *Biochem. Int.* Vol. 12, 1986, p. 889.

27 F. Sanger, D. Nicklen, and A.R. Coulson. "DNA sequencing with chain-terminating inhibitors." *Proc. Natl. Acad. Sci. USA*, Vol. 74, 1977, p. 5463.

28 F. Sanger, and A.R. Coulson. "The use of thin acrylamide gels for DNA sequencing." *FEBS Lett.* Vol. 87, 1978, p. 107.

29 J. Devereux, P. Haeberli, and O. Smithies. "A comprehensive set of sequence analysis programs for the VAX." *Nucleic Acids Res.* Vol. 12, 1984, p. 387.

30 K. Reed, and D. Mann . "Rapid transfer of DNA from agarose gels to nylon membranes." *Nucl. Acid. Res.* Vol. 13, 1985, p. 7207.

31 W. Mamo, G. Fröman, A. Sundås, and T. Wadström. "Binding of fibronectin, fibrinogen and type II collagen to streptococci isolated from bovine mastitis." *Microbial Pathogen.* Vol. 2, 1987, p. 417.

32 S.K. Hollingshead, V.A. Fischetti, and J.R. Scott. "Complete nucleotide sequence of type 6M protein of the group A *Streptococcus*." *J. Biol. Chem.* Vol. 261, 1986, p. 1677.

33 B. Guss, M. Lindberg, and M. Uhlén. "The gene for staphylococcal protein A". In: Bacterial immunoglobulin-binding proteins (Boyle, M. D. P., ed.) Vol. 1, 1990, p. 29, Academic Press., New York.

34 V.A. Fischetti, V. Pancholi, and O. Schneewind. "Conservation of a hexapeptide sequence in the anchor region of surface proteins from Gram-positive cocci." *Mol. Microbiol.* Vol. 4, 1990, p. 1603.

35 P.-E. Lindgren, M. McGavin, C. Signäs, S. Gurusiddappa, M. Höök, and M. Lindberg. "The nucleotide sequences of two different genes coding for fibronectin binding proteins from *Streptococcus dysgalactiae* and identification of active sites." *Eur. J. Biochem.* Vol. 214, 1993, p. 819.

36 B. Nilsson, T. Moks, B. Jansson, L. Abrahamsén, A. Elmblad, , E. Holmgren, C. Henrichson, A. Jones, and M. Uhlén. "A synthetic IgG-binding domain based on staphylococcal protein A." *Protein Eng.* Vol. 1, 1987 , p. 107.

37 M.J. McGavin, G. Raucci, S. Gurusiddappa, and M. Höök. "Fibronectin binding determinants of *Staphylococcus aureus* fibronectin receptor." *J. Biol. Chem.* Vol. 266, 1991, p. 8343.

38 M.J. McGavin, S. Gurusiddappa, P.-E. Lindgren, M. Lindberg, G. Raucci, and M. Höök. "Fibronectin receptors from *Streptococcus dysgalactiae* and *Staphylococcus aureus*; Involvement of conserved residues in ligand binding."

39 P.-E. Lindgren, C. Signäs, L. Rantamäki, and M. Lindberg. "A fibronectin-binding protein from *Streptococcus equisimilis*: Characterization of the gene and identification of the binding domain." Submitted.

40 V. Pancholi, and V.A. Fischetti. "Isolation and characterization of the cell-associated region of group A streptococcal M6 protein." *J. Bacteriol.* Vol. 170, 1988, p. 2618.

41 L. Björck, U. Sjöbring, and W. Kastern. "Structure of peptostreptococcal protein L and identification of a repeated immunoglobulin light chain-binding domain". *J. Biol. Chem.* Vol. 267, 1992, p. 12820.

42 L.-O. Hedén, E. Fritz, and G. Lindahl. "Molecular characterization of an IgA receptor from group B streptococci: Sequence of the gene, identification of a proline-rich region with unique structure and isolation of N-terminal fragments with IgA-binding capacity." *Eur. J. Immunol.* Vol. 21, 1991, p. 1481.

43 J.M. Kuypers and R.A. Proctor. "Reduced adherence to traumatized rat heart valves by a low-fibronectin-binding mutant of *Staphylococcus aureus*." *Infect. Immun.* Vol. 57, 1989, p. 2306.

44 L. Nelson, J.-I. Flock,, M. Höök, M. Lindberg, H.-P. Müller, . and T. Wadström. Adhesion in staphylococcal mastitis as vaccin components. In: New

45 U.K. Laemmli. "Cleavage of structural proteins during the assembly of the head of bacteriophage T4." *Nature* Vol. 227, 1970, p. 680.

TREATMENT OF STREPTOCOCCAL PHARYNGOTONSILLITIS: REPORTS ON PENICILLIN'S DEMISE ARE PREMATURE

M. Markowitz[1,2], M.A. Geber[1], and E.L. Kaplan[3]

Correspondence and reprint requests to: Milton Markowitz, M.D., Departmentof Pediatrics, University of Connecticut Health Center, FarmingtonConnecticut, 06030 - 1515

Introduction

Penicillin has been the drug of choice for the treatment of Group A beta hemolytic streptococcal infections for more than 40 years. After Denny and

1 Department of Pediatrics, University of Connecticut Health Center, Farmington, CT 06030-1515
2 Corresponding author
3 Department of Pediatrics, University of Minnesota School of Medicine

colleagues (1) showed that penicillin treatment of GABHS pharyngotonsillitis could prevent most first attacks of rheumatic fever, attention began to be focused on the accurate diagnosis and effective treatment of GABHS pharyngotonsillitis. Although a single injection of benzathine penicillin G became the "gold standard" for the treatment of GABHS pharyngotonsillitis and was once widely used to avoid problems with noncompliance, the pain and increased risk of an allergic reaction associated with this preparation led many physicians to prescribe penicillin orally (2). Currently, oral penicillin remains the treatment prescribed by the majority of physicians for GABHS pharyngotonsillitis. (3).

In recent years, however, it has been suggested that there has been an increasing incidence of bacteriologic treatment failures after a 10 day course of oral penicillin for GABHS pharyngitis (4,5). This report will review studies published between 1953 and 1993 which have used oral penicillin to treat GABHS pharyngotonsillitis, and compare the treatment failure rates among the studies with similar protocols for two time periods: 1953 to 1979 and 1980 to 1993. This chronologic grouping was selected because it was after 1980 that commentaries began to suggest that the efficacy of penicillin for GABHS pharyngotonsillitis might be diminishing (6,7).

Methods

A search of the literature between 1953 and 1993 was undertaken to locate studies of the use of oral penicillin for the treatment of GABHS pharyngotonsillitis. This review is limited to bacteriologic treatment failures. Virtually all of these studies involved a comparison of the efficacy of oral penicillin and other antibiotics, but this review will focus only on the efficacy of oral penicillin. In addition, only studies that required a 10 day course of a standard dose of oral penicillin and reported results as the percentage of patients with post-treatment persistence of GABHS in the pharynx were included.

These requirements were met by 65 studies which then were analyzed further to determine which fulfilled the following additional criteria established to assure the uniformity of the protocols: (1) virtually all subjects were in the pediatric age group, (2) all subjects had clinical evidence of acute pharyngitis, (3) all subjects had a positive throat culture for GABHS before treatment, (4) information was available to indicate how compliance for a full 10 day course was monitored, (5) one or more follow-up throat cultures were obtained within 60 days after completion of antibiotic therapy, and (6) there was appropriate use of serotyping (GABHS serotyping of the initial and follow-up cultures, with bacteriologic treatment failures defined as the presence of the same GABHS serotype in both the initial and follow-up throat cultures).

All of the above additional criteria were met by 32 of the 65 studies. Another group of 19 studies met all the additional criteria except the serotyping criteria and were, therefore, analyzed separately. The remaining 14 studies were eliminated for the following reasons: in five, sufficient information regarding compliance was not available; in five, the majority of the subjects were adults; and in four, the majority of the subjects were believed to be streptococcal carriers at the onset of antibiotic therapy.

The data were analyzed using meta-analysis[8].

Results

Of the 32 studies which included appropriate use of GABHS serotyping, 18 were published between 1953 and 1979 and 14 between 1980 and 1993. In the 18 studies published between 1953-1979 (representing a total of 1607 treated patients), the bacteriologic treatment failure rates determined from cultures taken from 1 to 60 days posttherapy ranged from 6 to 22% (mean 14%; 95% C.I. of 12.0% - 16.3%) (9, 13-29). Follow-up cultures from 1052 patients obtained between 1 and 14 days posttherapy showed bacteriologic treatment failure rates of 3 to 19% (mean of 10.5%; 95% C.I. of 8.3%-13.3%) (14-17, 20, 21, 23, 25-27).

In the 14 studies published between 1980 to 1993 (representing a total of 1176 treated patients), the bacteriologic treatment failure rates determined from cultures taken from 1 to 60 days posttherapy ranged from 2 to 30% (mean of 16.9%; 95% C.I. of 12.9%-21.8%) (7,10,30-41). Follow-up cultures from 514 patients obtained between 1 to 14 days posttherapy showed bacteriologic treatment failure rates of 2 to 21% (mean of 12.0%; 95 C.I. of 7.9%-17.8%) (31-33,35,37,42).

There was no significant difference in the bacteriologic treatment failure rates for the studies published during the earlier and later periods whether the follow-up period considered for both groups of studies was 1 to 14 days posttherapy or 1 to 60 days posttherapy (see Table).

Of the 19 studies whose design either did not include serotyping or did not restrict the failure rate to posttreatment cultures with homologous GABHS serotypes, 11 were published between 1953 and 1979 and 8 between 1980 and 1993. The bacteriologic treatment failure rates for the 11 studies published between 1953 and 1979 (representing a total of 2144 treated patients) ranged from 10 to 36% (mean of 21.7%; 95% C.I. of 17.4%-26.7%)) when considering cultures obtained between 1 and 60 days posttherapy (8,43-52). Follow-up cultures from 1587 patients obtained between 1 and 14 days posttherapy showed bacteriologic treatment failure rates of 5 to 15% (mean of 12.1%; 95% C.I. of 8.8%-16.5%) (8,45,48-50,52).

The bacteriologic treatment failure rates for the 8 studies published between 1980 and 1993 (representing a total of 496 treated patients) ranged from 8 to 24% (mean of 19.1%; 95% C.I. of 14.9%-24.2%) when considering cultures obtained 1 to 60 days posttherapy (5,53-59). Follow-up cultures from 425 patients obtained between 1 and 14 days posttherapy showed bacteriologic treatment failure rates of 4 to 22% (mean of 13.0%; 95% C.I. of 7.3%-20.8%) (5,53,55-58).

There was no significant difference in the bacteriologic treatment failure rates for the studies published during the earlier and later periods whether the follow-up period considered for both groups of studies was 1 to 14 days posttherapy or 1 to 60 days posttherapy (see Table).

Discussion

Upon reviewing these reports published over the past 40 years, one is struck by the wide range of results. Much of this variability can be attributed to inconsistencies in study design. Among the variables in study design which can influence outcomes are 1) the clinical status of subjects at the time the treatment is initiated, 2) whether bacteriologic treatment failures are based solely on the inability to eliminate the infecting serotype of GABHS, 3) how well compliance is being monitored, and 4) the timing of follow-up throat cultures.

The substantial impact that the clinical status of the subjects at the time the treatment is initiated can have on the outcome is apparent in several studies in which as many as 75% of the subjects were asymptomatic and the bacteriologic failure rates ranged from 21 to 38% (60-62). The authors of these studies explained that these high treatment failure rates occurred, at least in part, because the majority of the patients were GABHS carriers at the time therapy was initiated. It is well known that penicillin is not very effective in eradicating GABHS from someone who has become a carrier (42), yet these high failure rates are the ones often cited by those who suggest that the effectiveness of penicillin for GABHS pharyngotonsillitis has diminished (4,5).

The most important measure of an antibiotic's effectiveness is how well it eradicates the initial infecting serotype of GABHS from the upper respiratory tract. When serotyping is not done, or is not applied appropriately, treatment failure rates are exaggerated by including non-typeable strains and serotypes other than the infecting strain. As many as 60% of the GABHS cultured from patients followed for 2 months posttherapy are different serotypes from the initial infecting strain (30). Isolation of a heterologous strain on a follow-up throat culture constitutes a new acquisition, whereas isolation of a homologous strain usually constitutes a bacteriologic treatment failure. In a study by Breese, (43) the bacteriologic treatment failure rate was only 4% when only homologous serotypes of GABHS in the posttherapy culture were included and 12% when, in addition, non-typeable strains were included. In about one quarter of the reports reviewed for this evaluation, serotyping either was not done or was not used appropriately. Not unexpectedly, the bacteriologic treatment failure rates in these studies for cultures obtained 1 to 60 days posttherapy, were significantly greater than in the studies in which serotyping was performed appropriately.

Any positive culture with an homologous strain of GABHS that occurs after a previously negative posttherapy culture is considered a bacteriologic treatment failure. However, positive cultures within 2 weeks of the completion of therapy are much more likely to represent true bacteriologic treatment failures than cultures obtained more that 2 weeks posttherapy (30). The positive cultures with homologous serotypes that occur more than 2 weeks after completion of therapy can result from the reacquisition of the original strain from others. In the studies with appropriate serotyping, the bacteriologic treatment failure rates at 1 to 14 days posttherapy were significantly lower than the bacteriologic treatment failure rates at 1 to 60 days posttherapy.

Differences in patient compliance can also have a substantial impact on the bacteriologic treatment failure rate, and how compliance is monitored is an important consideration. The more often any drug has to be taken each day, the poorer compliance is likely to be (63). Therefore, careful monitoring of compliance is particularly important in drug trials which require three doses of oral penicillin daily. For example, in one study a 12% treatment failure rate was observed in compliant patients compared with a 34% treatment failure rate in non-compilers (44). To compare studies which have been carefully monitored for compliance with those that have not, is, therefore, inappropriate.

Even when only patients with acute pharyngitis are included, compliance is carefully monitored, follow-up evaluations are performed within 14 days of the completion of therapy, and serotyping of all isolates of GABHS is performed and properly used to distinguish true treatment failures from new acquisitions, about 10% of patients with GABHS pharyngotonsillitis who are treated with oral penicillin will have bacteriologic treatment failures.

When all of the studies analyzed in this report are considered, there is no direct evidence that the effectiveness of oral penicillin in treating GABHS pharyngotonsillitis has diminished in recent years. There also is no indirect evidence that the effectiveness of oral penicillin in treating GABHS pharyngitis has diminished in recent years. If penicillin therapy had become less effective over time, one would have anticipated an increase in the incidence of treatment failures which would have resulted in increased transmission of GABHS and an increase in the incidence of GABHS pharyngotonsillitis; however, this has not occurred (764. Nor is there any evidence of an increase in suppurative complications of GABHS pharyngitis (e.g. peritonsillar abscess). There has been an increased frequency of severe invasive disease caused by GABHS, as well as of patients with toxic shock syndrome (65,66); however, these illnesses have largely occurred in previously healthy individuals who had not received penicillin before the onset of symptoms. Therefore, there is no evidence that a change in the effectiveness of penicillin was responsible for the increased incidence of severe invasive GABHS disease or streptococcal toxic shock syndrome.

There also has been a recent increase in non-suppurative complications of GABHS pharyngitis in the form of several focal outbreaks of rheumatic fever occurring between 1985 and 1990 (64). However, among the patients for whom prior treatment information was available, three-quarters had received no antibiotic treatment because their preceding GABHS upper respiratory tract infection was either asymptomatic or too mild to warrant medical attention. Among the small number of patients who were treated with antibiotics and who, nevertheless, developed rheumatic fever, there was no information on the type of antibiotic prescribed, the length of treatment, or the level of compliance (73). Thus, there is no evidence to indicate that a change in the effectiveness of penicillin was responsible for these outbreaks of rheumatic fever, or that penicillin is any less effective as a prophylactic agent for the prevention of recurrences of rheumatic fever than it had been.

We conclude that bacteriologic treatment failures with oral penicillin therapy for GABHS pharyngitis, identified within 2 months after the completion of therapy, occur approximately 10 to 15% of the time. Although the specific reasons for these bacteriologic treatment failures have yet to be determined, there has been no substantial increase in the failure rate in recent years. Oral penicillin is as effective today as it was 40 years ago in the treatment of GABHS pharyngitis, and it remains the drug of choice for the treatment of this common pediatric problem.

Table 1 Comparison of findings in studies 1953-79 and 1980-83

Time period	No. of studies	Total no. of patients	Mean bacteriologic failure rate (95% C.I.) days posttherapy 1-14	1-60
1953-79	18 *	1607	10.5% (8.3%-13.3%)	14.0% (12.0%-16.3%)
1980-93	14 *	1176	12.0% (7.9%-17.8%)	16.9% (12.9%-21.8%)
1953-79	11 **	2144	12.1% (8.8%-16.5%)	21.7% (17.4%-26.7%)
1980-93	8 **	496	13.0% (7.3%-20.8%)	19.1% (14.9% - 24.2%)

*** with GABHS serotyping (Ref. 7, 9, 10, 13-41)**

**** with no or incomplete GABHS serotyping (Ref. 5, 8, 43-59)**

References

1. Denny FW, Wannamaker LW, Brink WR, Rammelkamp CH Jr, Custer EA. Prevention of rheumatic fever; treatment of the preceding streptococci infection. JAMA 1950; 143:151-153.

2. Gerber MA, Markowitz M. Management of streptococcal pharyngitis reconsidered. Pediatr Inf Dis 1984; 4:518-26.

3. Nelson JD, McCracken GH Jr. Streptococcal infections. Pediatr Infect Dis J 1990; 16:11.

4. Pichichero ME, Margolis PA. A comparison of cephalosporins and penicillin in the treatment of group A beta-hemolytic streptococcal pharyngitis: A meta-analysis supporting the concept of microbial co-pathogenicity. Pediatr Infect Dis 1991; 10:275-281.

5. Disney, FA, Hanfling MJ, Hausinger SA. Loracarbef (Ly163892) vs. penicillin VK in the treatment of streptococcal pharyngitis and tonsillitis. Pediatr Infect Dis J 1992; 11:520-527.

6. Pichichero ME. The rising incidence of penicillin treatment failures in Group A streptococcal tonsillopharyngitis: an emerging role for the cephalosporins? Pediatr Infect Dis J 1991; 10:S50-5.

7. Tanz., RR, Shulman ST, Stroka PA, Marubio S, Brook I, Yogev R. Lack of influence of beta-lactamase-producing flora on recovery of group A. streptococci after treatment of acute pharyngitis. J Pediatr 1990; 117:859-863.

8. Breese BB, Disney FA, Talpey WB. Triacetyloleandomycin-- a substitute for penicillin G. Am J Dis Child 1961; 101:423-428.

9. Stillerman M, Bernstein SH, Smith ML, Gittelson SB, Karelitz S. Antibiotics in the treatment of beta-hemolytic streptococcal pharyngitis: factors influencing the results. Pediatr 1960; 25:27-28.

10. Schwartz RH, Wientzen F, Pedreria F, Feroli EJ, Mella GW, Guandolo VL. Penicillin V for group A streptococcal pharyngitis tonsillitis. JAMA 1981; 264:1790-1795.

11. Kaplan EL. Benzathine penicillin G for treatment of group A streptococcal pharyngitis: a reappraisal in 1985. Pediatr Inf Dis J 1985; 4:592-596.

12. Der Simonian R, Lavid, N. Meta-analysis in clinical trials. Controlled Clinical Trials 1986; 7:177-188.

13. Stillerman M, Bernstein SH, Smith ML, Gorvoy JD. Erythromycin propionate and potassium penicillin in the treatment of group A streptococcal pharyngitis. Pediatr 1963; 31:22-28.

14. Stillerman M, Bernstein SH. Streptococcal pharyngitis therapy. Am J Dis Child 1964; 107:35-45.

15. Markowitz M, Kramer I, Goldstein E, et al. Persistence of group A streptococci as related to penicillinase-producing staphylococci: comparison of penicillin V potassium and sodium nafcillin. J Pediatr 1967; 71:132-7.

16. Rosenstein BJ, Markowitz M, Goldstein E, et al. Factors involved in treatment failures following oral penicillin therapy of streptococcal pharyngitis. J Pediatr 1968; 73:513-520.

17. Aronovitz GH, Morgan DL, Spitzer TQ. Streptococcal infections in pediatric patients. Comparison of dicloxacillin with phenoxymethol penicillin. Am J Dis Child 1968; 116:66-69.

18. Stillerman M, Isenberg HD. Streptococcal pharyngitis therapy: comparison of cyclacillin, cephalexin, and potassium penicillin V Antimicrob Agents Chemother 1970; 270-276.

19. Stillerman M. Comparison of cephaloglycin and penicillin in streptococcal pharyngitis. Clin Pharmacol Ther 1970; 11:205-212.

20. Howie VM, JH Ploussard. Treatment of group A streptococcal pharyngitis in children. Am J Dis Child 1971, 121:477-480.

21. Stillerman M, Isenberg HD, Moody M. Streptococcal pharyngitis therapy: comparison of cephalexin, phenoxymethyl penicillin, and ampicillin. Am J Dis Child 1972; 123:457-61.

22. Howie VM, Ploussard JH. Compliance dose-response relationships in streptococcal pharyngitis. Am J Dis Child 1972; 123:18-25.

23. Stillerman M. Isenberg HD, Facklam RR. Streptococcal pharyngitis therapy: comparison of clindamycin palmitate and potassium phenoxymethyl penicillin. Antimicrob Agents Chemother 1973; 4:514-520.

24. Lester RL, Howie VM, Ploussard JH. Treatment of streptococcal pharyngitis with different antibiotic regimens. Clin Pediatr 1974; 13:239-242.

25. Shapera RM, Hable KA, Matsen JM. Erythromycin therapy twice daily for streptococcal pharyngitis. Controlled comparison with erythromycin or penicillin phenoxylmethel four times daily or penicillin G benzathine. JAMA 1973; 226:531-535.

26. Matsen, JM, Torstenson O, Siegel SE, and Bacaner H. Use of available dosage forms of cephalexin in clinical comparison with penoxymethyl penicillin and benzathine penicillin in the treatment of streptococcal pharyngitis in children. Antimicrob Agents Chemother 1974; 6:501-506.

27. Stillerman M, Isenberg HD, Facklam RR. Treatment of pharyngitis associated with group A streptococcus: comparison of amoxicillin and potassium phenoxymethyl penicillin. Antimicrob Agents Chemother 1973; 4:514-520.

28. Breese BB, Disney FA, Green JL, Talpey W. The treatment of beta-hemolytic streptococcal pharyngitis. Comparison of amoxicillin, erythromycin estolate and penicillin V. Clin Pediatr 1977; 16:460-463.

29. Disney FA, Breese BB, Francis AB, Green JL, Talpey WB. The use of cefaclor in the treatment of beta-hemolytic streptococcal throat infections in children. Postgrad Med J 1979; 55 (suppl. 4):50-52.

30. Gerber MA, Randolph MF, DeMeo KK, Kaplan EL. Lack of impact of early antibiotic therapy for streptococcal pharyngitis on recurrence rates. J Pediatr 1990; 117:853-858.

31. Ginsburg CM, McCracken GM Jr, Crow SD, et al. A controlled comparative study of penicillin V and cefadroxil therapy of group A streptococcal tonsillopharyngitis. J Int Med Res 1980; 8 (supp,1):82-6.

32. Ginsburg CM, McCraken Jr GH, Steinberg JB, et al. Treatment of group A streptococcal pharyngitis in children: Results of a prospective, randomized study of four antimicrobial agents. Clin Pediatr 1982; 21:83-8.

33. Henness DM. A clinical experience with cefadroxil in upper respiratory tract infection. J Antimicrob Chemother 1982; 10 (Suppl B):125-35.

34. Gerber MA, Spadaccini LJ, Wright LL, Deutsch L, Kaplan EL. Twice-daily penicillin in the treatment of streptococcal pharyngitis. Am J Dis Child 1985; 139:1145-1148.

35. Stillerman M. Comparison of oral cephalosporins with penicillin therapy for group A streptococcal pharyngitis. Pediatr Infect Dis J 1986; 5:649-54.

36. Gerber MA, Randolph MF, Chanatry J, et al. Once daily therapy for streptococcal pharyngitis with cefadroxil. J Pediatr 1986; 109:531-537.

37. Pichichero ME, Disney FA, Aronovitz GH, Ginsburg C, Stillerman M. A multicenter randomized, single-blind evaluation of cefuroxime axetil and phenoxymethyl penicillin in the treatment of streptococcal pharyngitis. Clin Pediatr 1987; 26:453-458.

38. Pichichero ME, Disney FA, Aronowitz GH, Talpey WB, Green JL, Francis AB. Randomized, single-blind evaluation of cefadroxil and phenoxymethyl penicillin in the treatment of streptococcal pharyngitis. Antimicrob Agents Chemother 1987; 31:903-906.

39. Gerber MA, Randolph MF, Chanatry J, Wright LL, DeMeo K, Kaplan EL. Five vs ten days of penicillin V therapy for streptococcal pharyngitis. Am J Dis Child 1987; 141:224-7.

40. Gerber MA, Randolph MF, DeMeo K, Feder HM, Kaplan EL. Failure of once-Daily Penicillin V therapy for streptococcal pharyngitis. Am J Dis Child 1989; 143:153-155.

41. Disney RA, Dillon H, Blumer JL, et al. Cephalexin and penicillin in the treatment of Group A ß-hemolytic streptococcal throat infections. Am J Dis Child 1992; 146:1324-1327.

42. Kaplan EL, Gastanaduy AS, Huwe B. The role of the carrier in treatment failures after antibiotic therapy for group A streptococci in the upper r espiratory tract. J Lab Clin Med 1981; 98:326-335.

43. Breese BB, Disney FA, Talpey W,. Green JL, Tobin J. Streptococcal infections in children. Am J Dis Child 1974; 128:457-460.

44. Green JL, Roy SP, Charney E. Recurrence rate of streptococcal pharyngitis related to oral penicillin. J Pediatr 1969; 75:292-294

45. Breese BB: Treatment of beta-hemolytic streptococci infections in the home: relative value of available methods. JAMA 1953; 152:10-14.

46. Breese BB, Disney FA. A comparison of intra-muscular and oral benzathine penicillin G in the treatment of streptococcal infections in children. J Pediatr 1957; 51:157-163.

47. Breese BB, Disney FA. Penicillin in the treatment of streptococcal infections: a comparison of five different oral and one parental form. N Engl J Med. 1958; 259:57-62.

48. Breese BB, Disney FA, Talpey WB. Penicillin in streptococcal infections. Am J Dis Child 1965; 110:125-130.

49. Breese BB, Disney FA, Talpey WB. Beta-hemolytic streptococcal illness. A comparison of effectiveness of penicillin G, triacetyloleandomycin and erythromycin estolate. Am J Dis Child 1966; 111:128-132.

50. Breese BB, Disney FA, Talpey WB. Beta-hemolytic streptococcal illness. Comparison of lincomysin, ampicillin and potassium penicillin G in treatment. Am J Dis Child 1966; 112:21-27.

51. Colcher IS, Bass KW. Penicillin treatment of streptococcal pharyngitis. A comparison of schedules and the role of specific counseling. JAMA 1972; 222:657-657.

52. Randolph MF, Redys JJ, Cope JB, Morris KE. Streptococcal pharyngitis: posttreatment carrier prevalence and clinical relapse in children treated with clindamycin palmitate or phenoxymethyl penicillin. Clin Pediatr (Phila) 1975; 14:119-22.

53. Pichichero ME, Disney FA, Talpey WB, et al. Adverse and beneficial effects of immediate treatment of group A beta-hemolytic streptococcal pharyngitis with penicillin. Pediatr Infect Dis J 1987; 6:635-643.

54. Gooch WM, Higbee MD, Cocchetto DM, Evans EC. Cefuroxime axetil and penicillin V compared in the treatment of group A beta-hemolytic streptococcal pharyngitis. Clin Therap 1987; 9:670-677.

55. Stromberg A, Schwan A, Cors O. Five versus ten days treatment of group A streptococcal pharyngotonsillitis: A randomized controlled clinical trial with phenoxymethyl-penicillin and cefadroxil. Scand J Infect Dis 1988; 20:37-46.

56. Krober MS, Weir MR, Themelis NJ, Van Hamont JE. Optimal dosing interval for penicillin treatment of streptococcal pharyngitis. Clin Pediatr 1990; 29: 646-648.

57. Holm SE, Roos K, Stromberg A. A randomized study of treatment of streptococcal pharyngotonsillitis with cefadril or phenoxymethyl penicillin (penicillin V). Pediatr Infect Dis 1991; 10:S63-S71.

58. Brown RJ, Batts DH, Hughes GS, Greenwald CA. Comparison of oral cefpodoxine proxetil and penicillin V potassium in the treatment of group A streptococcal pharyngitis/tonsillitis. Clin Therap 1991; 13:579-587.

59. Ramet J. Pierard D, Vanderburghe P, De Boeck K. Comparative study of cefetamet pivoxil and penicillin V in the treatment of group A beta-hemolytic streptococcal pharyngitis. Chemotherapy 1992; 38 (suppl. 2):33-37.

60. Gastanaduy AS, Huwe BB, Kaplan EL, et al. Failure of penicillin to eradicate group A streptococci during an outbreak of pharyngitis. Lancet 1980; 2:498-502.

61. Smith TD, Huskins C, Kim KS, et al. Efficacy of B-lactamase-resistant penicillin and influence of penicillin tolerance in eradicating streptococci from the pharynx after failure of penicillin therapy for group A streptococcal pharyngitis. J Pediatr 1987; 110:777-782.

62. Kaplan EL, Johnson UR. Eradication of group A streptococci from the upper respiratory tract by amoxicillin with clavulate after oral penicillin treatment failure. J Pediatr 1988; 113:400-403.

63. Ayd FL. Single daily dose of antidepressants. JAMA 1974; 230:263-264.

64. Gerber MA, Markowitz M. Return of rheumatic fever to USA. Progress in Cardiology 1990; 3:177-86.

65. Shulman ST. Invasive and toxin-related diseases caused by group A streptococci. Pediatr Infect Dis J 1991; 10:528-531.

66. Stevens DL, Tanner MH, Winship J, et al. Severe group A streptococcal infections associated with a toxic shock-like syndrome and scarlet fever toxin A. N Engl J Med 1989; 321:1-7.

THE MECHANISMS OF LETHAL SYNERGISM OF MIXED INFECTION (INFLUENZA VIRUS AND GBS) AND ITS PREVENTION

R. Poljak[1,2], L. Chetverikova[1], K. Grabovskaya[1], I. Ivanova[1], T. Kramskaya[1], N. Shidlovskaya[1] and A. Totolian[1]

SUMMARY

Lethal synergism during mixed virus-bacterial infection (influenza A virus + Group B Streptococcus (GBS) in mice) has been shown to be caused by different mechanisms. Lethal synergism has the cyclic duration, closely connected with expression level of the virus polypeptides in the target-tissues. Peaks of lethal synergism coincide either with the peaks of maximum presentation of influenza virus antigen in the lungs or with the maximum level of cell cytotoxicity in the case of pathogenic GBS.

Disturbance of immunologic prevention during viral reinfection after initial mixed infection has been revealed and described. It was obvious that the level of antibodies circulating in the blood did not reflect the degree of prevention from the reinfection, which reduced, probably due to deficiency of formation of antigen-reactive clones of immunological memory limphocytes.

Present work substantiates the searching for preventive drugs from the development of severe forms of mixed infection. These kind of drugs can be selected among correctors of metabolic homeostasis.

1 Institute for Experimental Medicine Russia Academy of Medical Science acad. Pavlov's str., St. Petersburg, 197376, Russia

2 Corresponding author

INTRODUCTION

The results of our previous investigations of molecular and cellular pathogenic mechanisms of monoviral (influenza) and mixed (influenza and GBS) allowed to conclude that: 1) the central part of pathogenesis of viral and some forms of mixed infection is virus-specific modification of cell membranes; 2) the lethal forms of infection is accompanied by noncompensated activation of lipoperoxidation followed by damages of neurohumoral regulation and detoxication systems [1, 8, 9, 13].

It was also determined that even during lethal forms of monoviral or mixed infection the activation of general factors of specific or nonspecific defense aren't disturbed [1, 6, 15]. A level of activity of these reactions during infection had a ciclic character. Profile of cytotoxic activity curve highly corresponded to the accumulation of viral antigens in tissues as well as to the formation of target cells [5, 11, 15, 18].

During monoviral infection the possibility of correction of disease's severity by means of nonspecific drugs of general action (inhibitors of proteolysis, antioxidants, antihypoxants) was shown [9, 12, 14].

The main task of this investigation was search of the ways of the prevention of lethal forms caused by mixed infection. It was necessary to use immunologic and molecular approaches in order to study the possible reasons for lethal synergism [7].

MATERIALS AND METHODS

Pathogenic for mice strain of influenza virus A/PR/8/34 (H1N1), virulent 58/59-4M Fos 8 and avirulent 58/59-6S Fos 4 strains of GBS were used. Antibodies (AT) was determined in serum by the hemagglutination inhibition test. Antigens of influenza virus were detected in immuno-dot analysis with the modification [11]. The natural killers (NK) activity was studied by method [17]. In mathematically designed experiments [16] mice were intranosally inoculated by influenza virus A/PR/8/34 and virulent variant 58/54-4M Fos 8 of GBS in various doses (from 0,1 to 2,0 LD50). Antihypoxant Ac was intraperitoneally injected in doses 1, 5 and 9 mg/kg.

RESULTS AND DISCUSSION

Molecular and immunological mechanisms of lethal synergism during mixed infection

One of the main cause of the mixed infection aggravation is virus-specific modification of cell membranes. Built-in virus proteins are receptors for GBS and by that can caused transformation of a nonlethal infection to a lethal one [3-5]. The new approach was used for additional evidence of the role of virus polypeptides expressed on the membrane of mouse lung cells. Every day the sublethal influenza A virus infection was followed by additional inoculation of GBS. Two GBS strains were used: virulent Fos 8 and avirulent Fos 4. For monobacterial infection mouse mortality didn't exceed 10%. The different periods between two inoculations (from 1 to 10 days) were used. Animals (10 mice for each groups) were observed for 14 days after bacterial superinfection and lethality index was counted for each group.

In control group during monoviral infection the levels of infectious virus and viral antigens in lungs as well as NK activity were determinated daily. It allowed us to estimate the influence of value of these indicators (at the moment of inoculation of GBS) on the mixed infection. It was proved [5] that the virus reproduction in lungs during monoviral infection doesn't differ from it during mixed infection. The leading role of virus adhesion in increasing of virulence of GBS Fos 4 was proved. Intranosal inoculation of 107 CFU/ml of GBS didn't lead to its accumulation in the lungs (fig.1a). Virus reproduction was the same in case of monoviral and mixed infection (fig.1b). However inoculation of GBS in a day after virus infection beginning (10 lg EID) resulted in marked stimulation of bacterial growth. Intensive GBS reproduction led to transformation of a nonlethal form of infection to lethal one (80-100% mortality was observed in 5 days).

The content of viral antigens in lungs has a significant distinction (fig.2). Beginning with 5th or 6th days content of antigens in target-tissue has a steep rise in comparison with monoviral infection. It is followed by activation of NK activity.

It made us study whether the level of lethal synergism during GBS superinfection is changing accordingly the ciclic fluctuations of viral antigens expression on the cell membranes (fig.3b). It was shown that index of lethality also changes in a ciclic manner. Maximum of lethality for group of mice infected by virus plus avirulent Fos 4 strain coincides with maximum of viral antigens in the lung tissue. Increased mortality was detected on the 2nd, 5-6th and 8th days. In the other case of (virus + virulence Fos 8 strain) days of increased mortality didn't coincide with maximum of viral antigens content. And on contrary these mortality peaks were observed simultaneously with the maximum of NK activity on the 3d, 5-7th, 9th and 12th days. These data speak in favorite of different mechanisms of lethal synergism for the strains of GBS used. For the Fos 4 strain it is, evidently, connected with the increased presentation of viral polypeptides on the surface of lung cells; and for the Fos 8 strain - with additional opening of the cells surface receptors during intensive damages of virus-specific target-cells by effector-cells.

These data show that the pathogenesis of virus-bacterial infection depends on property of the certain pair of infected agents. In each case lethal synergism connected with different molecular and immunological mechanisms but every time closely related with viral proteins expression on the target-organ cell surface and cytotoxic activation. Mechanisms of lethal synergism must be further studied in details.

Impairment of resistance to the virus reinfection after primary virus-bacterial infection

Our recent studies had shown that both GBS infection [2, 5] and stress [8] induced aggravation of the experimental influenza in mice.

It was shown that influenza virus infection developed

under stress condition accompanied by reduction of resistance to repeated infection by homological influenza virus. It depended on the defects of immunological memory lymphocytes clones defects [13].

The influence of bacterial superinfection on resistance to the secondary infection with the same virus was studied in this investigation. The value of it was estimated by a level of virus reproduction after reinfection (table 1). It was shown that bacterial superinfection independently to the virulence of the strain and to the time of its inoculation leads to the disorders in the mouse resistance to viral reinfection. Through a month these mice weren't protected to reinfection in spite of the animals primary infected only by influenza virus.

Table 1. Resistance of mice to virus reinfection after mono- or mixed (influenza + GBS) infection

Type of infection	Time for administration of GBS in respect of beginning of virus infection	Titer of antiinfluenza AT before reinfection	Level of virus reproduction (lg) in 2 days after reinfection*
intact (control)			6,5
Monoviral (3,5 lg EID)		160	<1,0
mixed A/PR/8+GBS Fos 8 in dose 10 CFU	1 day before	160	3,5
"	simultaneously	80	5,5
"	3 day after	80	5,5
Monoviral (3,0 lg EID)		320	<1,0
mixed A/PR/8+GBS Fos 4 in dose 10 CFU	3 day before	160	5,0
"	3 day after	320	5,5

* - A/PR/8 in dose 6,0 lg EID

Reproduction of virus in the lungs of superinfected mice during secondary virus infection was practically the same as that of intact mice. Meanwhile antiviral antibodies synthesis was the same in all groups of animals.

Data obtained show the common features of these damages under influence of different agents: stress [8, 9, 13] or GBS superinfection. It is significant that the level of circulating anti-influenza antibodies doesn't reflect the degree of resistance to reinfection. At the same time it's very important that mixed virus-bacterial infection not only accompanied by its transformation into more severe disease but also leads to the defects of immune memory formation. It happens on the background of high values of the humoral immune response.

Usage of metabolic correctors as a drug for prevention of severe forms of virus-bacterial infections

Taking in account the importance of mixed infection problem as well as variety of lethal synergic mechanisms the search of new nontraditional protective drugs was done. In previous experiments it was shown that using of nonselective drugs, returning the metabolic homeostasis to normal provides the defence from the severe form of influenza virus infection. Such drugs can be selected among the inhibitors of proteolysis, antioxidants and antihypoxants [1, 12, 14].

Based on the idea of damaging mechanisms identity of different pathogenic factors we studied the action of antihypoxant (Ac) as a prophylactic remedy during mixed infection.

By means of mathematically designed experiments the action

of Ac was studied [1, 16]. It was shown (table 2) that the single intraperitoneal injection in dose 9 mg/kg one day before virus infection led to 30%-50% less mortality during mixed infection (influenza + GBS Fos 8). Since the protective action of Ac during monoviral infection results in limitation of noncompencated activation of lipoperoxidationn, it may be supposed that in the case of mixed infection antihypoxant provides correction of metabolic homeostasis. These data concerning possibility of influence on the severity of pathogenic process during mixed infection let us propose substitute new nontraditional approach of in searching of anti-infective remedies among the correctors of metabolic homeostasis.

Table 2. Prophylactic effect of antihypoxant Ac for mixed virus-bacterial infection in mice

Injection of Ac (9 mg/kg)	Mortality of mice (%) during 14 days after simultaneous infection by influenza virus (0,5 LD50) and GBS (1,0 LD50) in experiments No			
	1	2	3	4
-	90	100	100	10
+	60	50	60	70

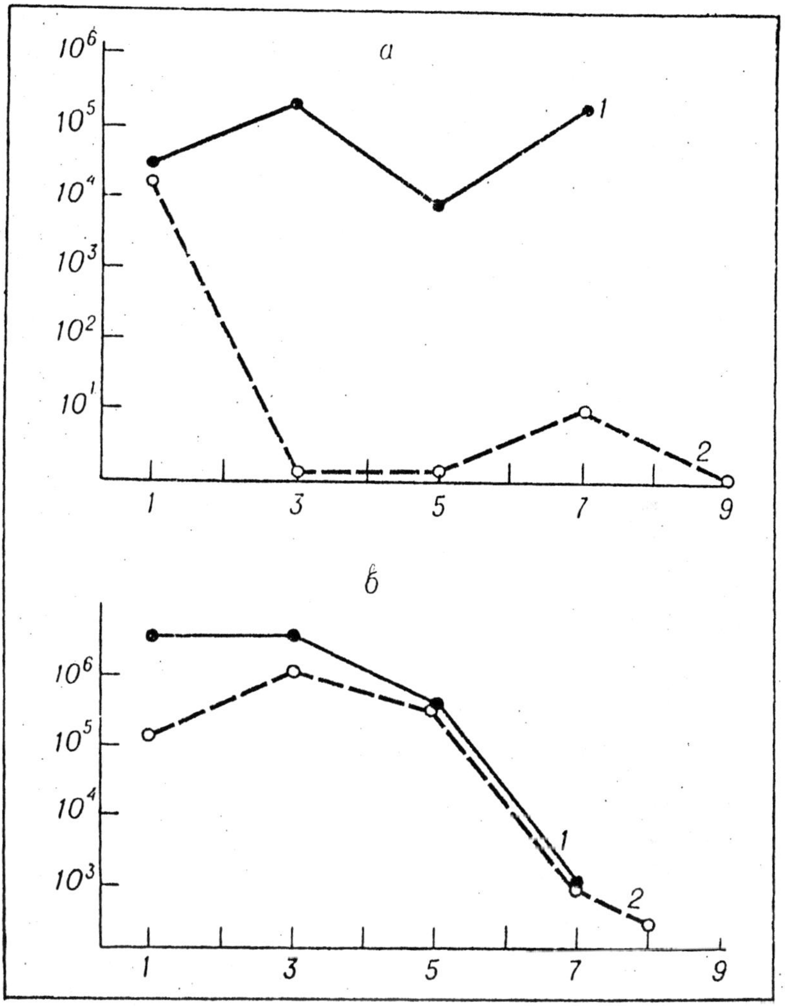

Fig.1 Accumulation of infection agents in mouse lungs during monoviral and mixed infection

a - GBS Fos 4

b - A/PR/8/34

1) - mixed infection; 2) - monoviral infection

X axis - days after infection; Y axis - level of GBS (CFU) and A/PR/8/34 (IED)

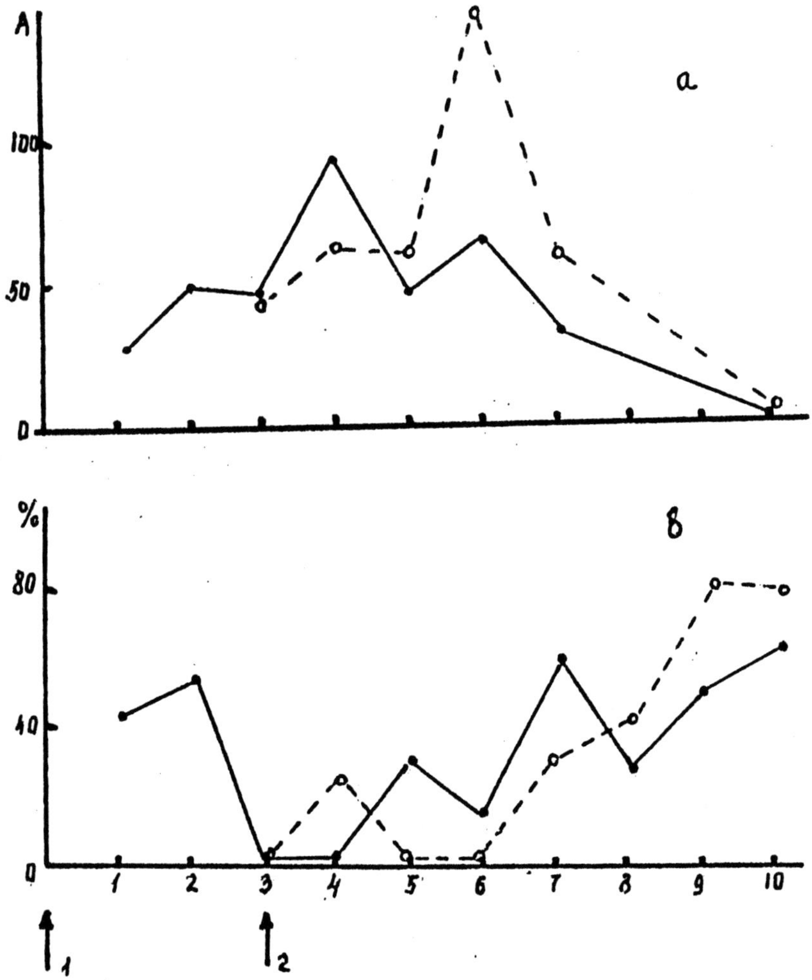

Fig.2 Change of expression of influenza virus antigen in the lungs and NK activity in the spleen after mixed infection

a - level of virus antigens

b - NK activity

_____ monovirus infection

- - - mixed infection

Infection: 1 - A/PR/8/34

2 - GBS Fos 8

A - % - NK activity

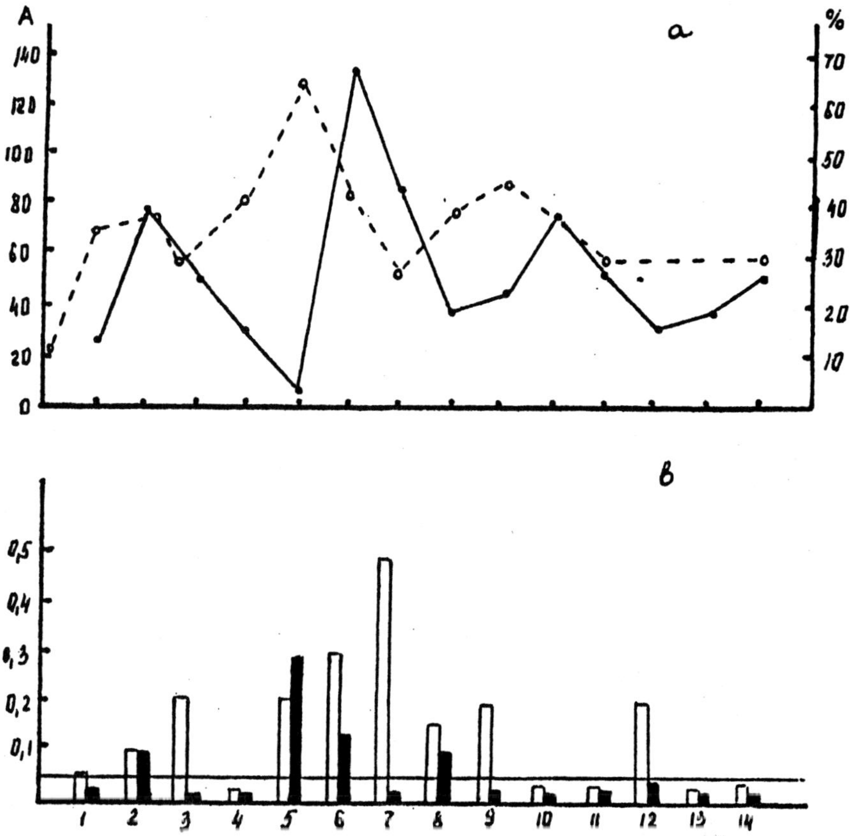

Fig.3 Lethal synergism of influenza A virus and GBS in mice.

X axis - days of vital infection (4 lg EID)

a - (- - -) viral antigen in lungs A - (absorbancy at 633 nm x 100)

(______) NK activity (in %)

b - index of lethalyty (on the 14th day of GBS infection) after GBS inoculationon the corresponding day of viral infection

- Fos 8 - 6 lg CFU/50 mkl
- Fos 4 - 8 lg CFU/50 mkl

______ - horizontal line maximal lethality index of monoinfections

REFERENCES

1. Chetverikova L.K., Kildivatov I., Inozemtseva L.I. et al. Factors of antiviral resistance in the pathogenesis influensa infection in mice. Vestnik Acad. Nauk SSSR, 1989, N11, P.63-68.

2. Grabovskaya K.B. Postinfluenza streptococcal infection in vitro and in vivo. In: Strategy of infection in the host body. Leningrad., 1987, p.122-131.

3.Grabovskaya K., Bulgakova T., Dubrovina T. et al. (1986). The influenza a virus as an inductor of tissue susceptibility to infection by group B streptococci. Bacteria and host. Eds. Ryc. M., Frank I., Prague, P.47-51.

4. Dubrovina T., Grabovskaya K., Zueva O. et al. (1986). Virus-specific modification of cell membranes as a factor of group B streptococcus adhesion. Vestn. Akad. med. sci. USSR, N7, P.50-57. (in Russ.).

5. Dubrovina T., Grabobskaya K., Ivanova I. et al. (1989). Fatal synergism of virus-bacterial infection (model: influenza - streptococcus). Vestn. Akad. med. sci. USSR, N11, P.17-22. (in Russ.).

6. Dubrovina T., Leontieva G., Polyak R. General and specific events in realization of defense reactions in viral infection (facts and hypothesises). Vestn. Akad. med. sci. USSR, 1991, N4, P.3-7.

7. Editorial. How does influenza virus pave the way for bacteria? The Lancet, 1982, P.485-486.

8. Frolov B.A., Chetverikova L.K., Kramskaya T.A. Polyak R.Ya. Influence of immobilization stress on the course of experimental influenza infection and the process of interferon production. J. microbiol., epidem., immunolbiol., 1986, N3,

p.53-56.

9. Frolov B., Polyak R., Smagin G., Niconorov A. et al. Pathogenic mechanisms of stress-induced experimental influenza aggravation and its prevention. Vestn. Akad. med. sci. USSR, 1991, N4, P.50-54.

10. Haffejce Y.E., Bhana R.H., Coovadia Y.M. et al. Neonatal group B streptococcal infectious in Indian (Asian) babies in South Africa. J. of Infection, 1991, V.22, P.225-231.

11. Ivanova I., Dubrovina T., Meringova L. at al. (1989). Determination of viral antigen in the lungs and the spleen in influenza virus infected mice. Mol. Genet. Microbiol., virol., N5, P.39-45. (in Russ.).

12. Inozemtseva L.I., Chetverikova L.K. The role of lipid peroxidation in poststressor aggravation of influenza infection and its ionol correction. Vestnik Acad. Nauk SSSR, 1991, N4, P.54-57.

13. Kramskaya T.A., Chetverikova L.K., Frolov B.A., Polyak R.Ya. Poststressor loss of mice resistance to repeated infection by homological influenza virus. Vestnic AMN USSR, 1991, N4, p.47-50.

14. Lozitsky V.P., Polyak R. The role of proteolysis in the reproduction of animal and human viruses and antiviral activity of protease inhibitors. Adv. mod. Biology., 1982, V.93, P.352-362.

15. Polyak R., Voitsekhovsky B.,Dubrovina T. et al.(1991). Biological balance in the body during virus infection of various severity (in a model of influenza). Vestn. Akad. med. sci. USSR, N4, P.34-40 (in Russ.).

16. Razorenov G.I. Automation of an analysis of a mathematical models in epidemiological process. In:

Mathematical methods in epistemology and microbiology, Leningrad, 1979, P.46-53.

17. Rykova M., Spirande I., Zergenidze M. at al. (1981). New highsensitive technics for nátural killer cells test. Immunology, N3, P.88-90. (in Russ.).

18. Sanford B., Shelocov A., Ramsay M. Bacterial adherence to virus - infected cells: a cell culture model a bacterial superinfection. J. Infect. Dic., 1978, V.137, P.176.

MECHANISMS FOR VARIATION IN THE M PROTEIN OF STREPTOCOCCUS PYOGENES

W.A. Relf, D.R. Martin[1], B.N. Manjula[2], and K.S. Sriprakash[3]

ABSTRACT

The highly variant M protein of group A streptococci (GAS) performs a vital role in establishing streptococcal infection. Knowledge on the structure and organisation of the genes encoding M proteins (*emms*) is restricted to a subset of well characterised M molecules. The evolutionary relationships between DNA and translated peptide sequences representing many serologically distinct M proteins were examined. Certain *emms* shared similarities within the variable regions and were classified into families of M proteins. The study of closely related sequences proved useful for examining events resulting in the generation of M protein diversity. Point mutation, insertion and deletion of small and large repeated sequences, and compensatory frameshift mutation are suggested as contributing to the process.

1 New Zealand Communicable Disease Centre, Porirua, New Zealand
2 The Rockefeller University, NY USA
3 Menzies School of Health Research, PO Box 41096, Casuarina, Australia 0811 Phone (61-89-228612); Fax(61-89-275187); Email(sri@menzies.su.edu.au); (corresponding author)

1. INTRODUCTION

The most important and widely studied virulence factor for GAS is the dimeric alpha-helical coiled-coil M protein protruding from the bacterial surface [1,2]. Antibodies to the M protein are type-specific [3], and promote opsonisation of GAS, thereby being protective [4,5]. The type-specific nature of the M protein is the basis of serotyping for GAS, by which more than 80 M serotypes have been designated. A number of GAS do not react with available sera and are designated as M nontypable (MNT).

Only a small number of M proteins and their genes (*emms*) have been characterised [6-11]. The N terminal signal peptide and the C terminal half are conserved between M proteins. The N terminal region is highly variable between mature proteins and contains serotype-specific and protective epitopes [12-15]. Downstream from the extreme hypervariable N terminal region are repeat regions designated as A, B and C, which display decreasing variability between distinct M proteins [7].

Various mechanisms contributing to antigenic diversity in M proteins have been suggested. The presence of the ML (M-like) protein gene family in GAS is likely to have arisen by the duplication of a common ancestral gene [11]. *emm* and the genes for the ML proteins (collectively known as *emm*L), are located in tandem arrangement [16-20]. DNA encoding the conserved signal peptide and membrane anchor regions may be an ideal substrate for intergenic recombination between *emm*s and *emm*Ls [11,21]. Base substitutions and duplication/deletion of repetitive sequences can lead to diversity within an M type. Studies of multiple M1 *emms* suggest that base substitutions can lead to altered antibody recognition of opsonic epitopes [22].

Deletion/duplication of sequences encoding the A, B and C repeats via intragenic recombination has also been suggested. The generation of size mutants occurs at a high frequency *in vitro* [23]. The change in size was attributed to spontaneous duplications or deletions within the repeats via intragenic recombination [24]. Some size variants had altered in antigenic nature which modified the antiphagocytic activity of the M6 protein [25]. Duplication/deletion of inexact repeats may enhance the change in type-specific and opsonic epitopes. The effect of point mutations present in repeats may be accentuated by duplication [21].

Recently, a 21 base pair insertion was found in the extreme nonrepetitive N terminus of isolates closely related to M1 [22]. The mechanism for the insertion/deletion in the nonrepetitive region is not known.

We have analysed a large number of sequences that potentially encode the highly variable regions of M proteins. Such a large database is useful for uncovering the genetic events that may contribute to M protein diversity. One of these was the novel frameshift and compensatory frameshift mutations.

2. METHODS

2.1 PCR and sequencing

The methods used to amplify *emm*L DNA, cloning and sequencing have been described elsewhere [26,27].

2.2 DNA and amino acid analysis

A suite of software programs offered by the Australian National Genome Information Service was used for all DNA and protein analyses. The programs include Staden programs and those offered by Genetics Computer Group, Wisconsin (GCG). The secondary peptide structure was predicted using PEPTIDESTRUCTURE (GCG) [28]. Data bases of DNA and amino acid sequences (without the conserved signal peptide region and C terminal region beginning with the C repeat), representing the known M and ML sequences [6-11,16,18-20,27,29-37] were used to compare each member with the others using MATCH [38], and GAP (GCG) [39]. Multiple sequence alignment and dendrogrammatic trees were generated using PILEUP (GCG) [40].

3. RESULTS

3.1 PCR and sequencing

The partial nucleotide sequences corresponding to the N terminal region from reference strains and isolates of M1, M5, M12, M24, M49 and M57 were 95-100% homologous to the published *emm* sequences [6,8-11,31]. The sequences from the reference strains and isolate of M2, M3 and M4 were 99-100% homologous to the *emm*L sequences of *emm*L2.1 [33] and *emm*L2A [32], *emm*L3 [36], and *arp*4 [29], respectively. The translated peptide sequence from the reference strain for M60 was 100% homologous to the published partial peptide sequence for Arp60 [30]. The translated peptide sequences for a reference strain and an isolate of M30 were highly homologous to the sequence from PepM30 (Figure 1). These results show selective amplification of *emm* and *emm*L sequences by the PCR method described earlier [26].

A total of 81 reference strains and isolates were studied. Where the peptide sequence data are unavailable, the sequences are designated as ML. STs were proposed for distinct sequences from the MNT isolates [27]. Thirty-seven different *emm*/*emm*Ls were distinguished and 28 of these have not been previously reported. Of the 32 MNT isolates studied, 15 distinct *emm*Ls were identified. Therefore the MNT isolates represent a heterogeneous group. Table 1 shows a summary of the structural features of the M and ML proteins studied. The length of the N terminal nonhelical portion varied from 0-61 residues. Repeated sequences were present in only some sequences and the number and length varied considerably (Table 1).

3.2 Comparison of *emm*L/ML sequences

*emm*Ls representating each sequence type were examined for similarity over short stretches of DNA. The analysis revealed five families of sequences and were designated as the 525380 family, ML2 family, TIN family, ML60 family and the M5 family. The homology between the members of 525380 family and between members of the ML2 family was present over the entire sequences examined. Localised regions of homology were observed within the TIN family members, the ML60 family members and the M5 family members.

High overall homology between both the nucleotide and translated amino acid sequences was observed for the members of the 525380 family (ML41, ML52, ML53, ML80, ST88/25, ST156, STBSB75, STNS5, STNS27) [41]. Homology was present within the region corresponding to the extreme N terminus and increased as the extreme N terminus was departured. We have previously suggested that this homology may reflect the crossreactions observed in serotyping between GAS of the M types represented in the 525380 family [41].

Alignments for members of the ML2 family showed considerable homology at the nucleotide and amino acid level within the N terminal hypervariable and variable regions. *emm*L2.1 and *emm*LBSA10 were 85% homologous, *emm*L2.1 and *emm*LNS14X were 74% homologous, and *emm*LBSA10 and *emm*LNS14X were 66% homologous. There was 56% overall homology between the three nucleotide sequences.

In contrast to the 525380 and ML2 families, nucleotide similarity for the TIN family was localised to specific regions of DNA (Figure 2a). *emm*L90/85 and *emm*L88/544 displayed 96% homology over 119 bases, *emm*L90/85 and *emm*L13 displayed 93% homology over 28 bases, and *emm*L90/85 and *emm*L1293 displayed 79% homology over 29 bases. *emm*L13 and *emm*L1293 displayed 82% homology over 130 bases. *emm*L88/544 was 81% homologous to *emm*L11 over 141 bases, 60% homologous to *emm*L88/372 over 132 bases, and 86% homologous to *emm*L59 over 128 bases. *emm*L11 was 69% homologous to *emm*L8/372 over 162 bases, and 81% homologous to *emm*L59 over 130 bases. *emm*L88/372 was 61% homologous to *emm*L59 over 98 bases. The translated sequences for ML13 and ST1293 contain A repeats homologous to ST90/85 (Figure 2a). ST88/544, ML11 and ST88/372 share homology with the region corresponding to downstream from the A repeat region in ST90/85.

The nucleotide similarity for the ML60 family is shown in Figure 2b. *arp*60 and *arp*4 displayed 88% homology over 95 bases, *arp*60 and *emm*L8 displayed 81% homology over 132 bases, and *arp*60 and *emm*L22 displayed 86% homology over 77 bases. *arp*4 was 84% homologous to *emm*L8 over 80 bases, and *emm*L8 was 59% homologous to *emm*L22 over 34 bases. There was little homology between *arp*4 and *emm*L22. The translated amino acid sequences for Arp60 and ML22 are similar within, and downstream from the A repeat region (Figure 2b). Arp4 and ML8 have amino acid homology with Arp60, in the region downstream from the A repeat.

The nucleotide homology within the M5 family, was in discrete regions, except for *emm*5 and *emm*6. There is 79% overall DNA homology between *emm*5 and *emm*6 corresponding from the N terminus to the beginning of the C repeat [7,9]. The other M5 family members had homology at the 5' end of the *emm*L sequence. *emm*5 has 62% homology with *emm*30 over 131 bases, 69% homology with *emm*L18 over 119 bases, and 71% homology with *emm*LNS1 over 178 bases. *emm*6 has 61% homology with *emm*30 over 108 bases, 70% homology with *emm*L18 over 84 bases, 73% homology with *emm*LNS1 over 150 bases. *emm*30 has 69% homology with *emm*L18 over 130 bases, and 69% homology with *emm*LNS1 over 138 bases. *emm*L18 and *emm*LNS1 have 68% homology over 154 bases. All members of the family have amino acid homology within the region corresponding to the A repeat region of M5 (Figure 3). The homology between M5 and M6 extended though the B repeat region [9]. *emm*L18 showed a deletion of three nucleotides causing a loss of one amino acid.

The overall genetic relatedness between all reported ML sequences was plotted dendrographically (Figure 4). A similar display was found when comparing nucleotide sequences (data not shown). As expected, the families clustered together. ML52 was closest to ST156, and ML53 was closest to ML80. In the ML2 family, ML2.1 and STBSA10 were more closely related to each other, than to STNS14X. The dendrogram also reflected the similarity within the MLs of the TIN family. ST550 and Arp60 were most closely related within the ML60 family. M5 and M6 were most closely aligned within the M5 family, and M1, ML14, ML17 and M24 also showed relatedness to the

M5 family members (Figure 4).

3.3 Mechanisms of antigenic diversity

3.3.1 Point mutations

The alignment of multiple sequences derived from the same M type demonstrated the high frequency of base substitutions. *emm* from a reference type strain, 2031 was 100% homologous to the published *emm*1.0 [8,22]. Two *emms* from M1 GAS isolated from the Northern Territory of Australia were 100% homolous to *emm*1.1, derived from an M1 variant [22]. Due to base substitutions, M1.1 differed to M1.0 by 13 amino acids in the N terminal nonrepetitive region [22]. The changes resulted in the loss of an M1 opsonic epitope despite no change in M type. Likewise, base substitutions led to four amino acid differences in the N terminal nonrepetitive region for the reference type strain for M30 and an isolate of M30 (Figure 5a). It is not known if this reflects an alteration in opsonic epitopes in this region.

3.3.2 Changes in the repeat regions

The number of A repeats varied between isolates of ST88/25. Isolate 90/86 contained a duplication of a 14 amino acid A repeat region (Figure 5c). An M30 reference type strain 2054, lacks the first seven amino acids of a 14 amino acid A repeat (Figure 5b). Four complete A repeats were determined for M30 isolate 1GL251 (Figure 5b), and base substitutions appear to be responsible for the variation in amino acid sequence between the tandem repeats.

Variation with the B repeat region common to all members of the 525380 family was also observed. ML80, and ML52 have two 16 amino acid repeats while ST88/25 and M41 have a single B repeat [20,41]. ML53 lacks three amino acids in the first of two B repeats.

3.3.3 Frameshift mutation

Pairwise nucleotide alignments corresponding to the N terminal nonrepetitive regions of some members of the 525380 family revealed a single base deletion/addition causing a shift in frame for the amino acid sequence. After a short stretch of nucleotides, a compensatory frameshift corrected the reading frame [41]. High DNA homology was maintained throughout this region, while the frameshift/compensatory frameshift ensured a short stretch of novel amino acids. This was noted when comparing ML52 and ML53, ML52 and ML80, ML53 and ST88/25, and ML80 and ST88/25 [41]. Preliminary data also suggests the occurence of a frameshift and subsequent compensatory frameshift within the TIN family (data not shown).

4. DISCUSSION

Five ML families, 525380 family, ML2 family, ML5 family, ML60 family and TIN family are described. The families were designated based on DNA similarities in the N terminal sequences. The homology was present over most of the sequence for the 525380 family and the ML2 family, and was between specific sequences in the ML5 family, the ML60 family and the TIN family.

MLs from each family showed genetic relatedness within the extreme N terminal region, which has been previously deemed as highly variable between M types. These common regions within families may contribute to crossreactions in serotyping [42-44]. Similarities between MLs have been previously reported. EnnX and Enn4 share 86% homology in the N terminal half [19]. Many reports in the literature support the sequence similarity between M5 and M6. Various workers have noted crossreactions between M5 and M6 proteins [13,44-46]. Epitopes common to M5 and M6 have been identified within the N terminal half using polyclonal heterologous antisera [13], monoclonal antibodies [46], synthetic peptides[46] and DNA hybridisation [47]. In general, the dendrogram was useful for examining the overall relationships between many M and ML proteins and reflects the localised homology observed within families. More extensive sequencing towards the beginning of the conserved C repeat region may clarify the definition of the members of related ML proteins.

Differences in the number of repeated sequences was observed for MLs of M30 and ST88/25. In ST88/25, the outer boundaries of the repeats showed mismatches, with the core of the repeat remaining conserved. This feature has been previously described in M6 repeats [24]. Despite the difference in the number of A repeats in isolates of ST88/25s, they were assigned to the same ST.

Point mutations are important in ML diversity, and this is likely to contribute to gradual change in the N terminus [21]. Base substitutions were responsible for variation in amino acid sequences observed between multiple sequences of the same M type and those from closely related M types. This appears to be the most common form of mutation found in this study. The point mutation is a primary mechanism for variation in the repetitive and nonrepetitive regions of ML DNA [22,23]. In addition, the insertion of short nucleotide stretches may contribute to the variation in the nonrepetitive region [22]. We have also noted insertions of short nucleotide stretches within a family of ML proteins that vary in M type (data not shown).

Frameshift and compensatory frameshift mutations were found to occur in the N terminal nonrepeating regions, causing short stretches of unique residues flanked by amino acid homology. These stretches of unique sequences may represent possible candidates for type-specific and/or opsonic epitopes. The high degree of homology between the nucleotide sequences is unlikely to be explained by chance. The frameshift is very effective at altering the amino acid sequence. It modifies a limited region within the hypervariable region without any apparent effect on the helical structure at the N terminus. The compensatory frameshifts observed may reflect the increased selective pressure to maintain the coiled-coil structure of the protein.

Overall, the variation at the 5' end of *emms* corresponding to the N terminal hypervariable region could be attributed to a) preponderant base substitution mutations, b) insertions and deletions of short sequences without causing a disturbance in the reading frame and c) frameshift-compensatory mutations causing amino acid sequence divergence in the region limited by these events.

5. REFERENCES

1 B.N. Manjula and V.A. Fischetti; "Tropomyosin-like seven residue periodicity in three immunologically distinct streptococcal M proteins and its implications for the antiphagocytic property of the molecule"; *Journal of Experimental Medicine*, "Vol.151"; 1980; p.695.

ML Protein	Length of nonhelical region (amino acids)	Repeated sequences	
		number	length (amino acids)
M1	21	2	28
ML2	6	2	7
ML3/ML3R	8	-	
Arp4	16	-	
M5	10	6	7
ML8	14	-	
ML11	0	-	
M12	0	-	
ML13	25	-	
ML14	12	-	
ML17	11	2	14
ML22	21	-	
M24	9	-	
M30	11	2	14
M49	0	-	
ML52	10	2	16
ML53	1	2	9
ML55	61	-	
M57	21	2	33
ML59	14	-	
Arp60	0	3	7
ML75	0	-	
ML80	0	2	16
STBSA10	8	2	7
STBSB24	13	-	
STBSB75	0	-	
STDONALD	4	2	32
STNS1	7	5	7
STNS5	5	-	
STNS14X	6	2	11
STNS27	35	-	
ST88/25	19	-	
ST88/372	16	-	
ST88/544	20	-	
ST90/85	16	5	7
ST156	20	-	
ST1293	18	-	

Table 1. Summary of structural features of ML and M proteins derived from DNA sequencing in this study. Where underlined, the features of a reference strain are used as the prototype. The dash indicates that no repeat sequences were detected within the sequence available.

```
       Y  S  L  R  K  L  K  K  G  T  A  S  V  A  V  A  L  S
ctgcagtattcgcttagaaaattaaaaaaaggcactgcttcagtagcagtggctttgagt  60
 V  I  G  A  G  L  A  V  N  Q  T  E  V  S  A  T* I  I  T  R
gtaataggggcaggattagcggttaaccaaacagaagttagcgcaacaattattactcga 120
 A  I  A  K  D  P  D  K  L  K  E  I  A  E  N  F  E  V  Q  N
gctatagcaaaagatccagacaaattaaaagaaatagctgaaaatttgaggtacagaat 180
 H  E  L  T  R  K  N  G  E  L  T  Q  E  K  D  E  L  T  R  K
catgagttaacacgtaagaatggtgagttaacacaggagaaggatgagttaacacgtaag 240
 N  G  E  L  T  Q  E  K  D  E  L  T  H  E  N  G  E  L  T  Q
aatggtgagttaacacaggagaaggatgagttaacacatgaaaatggtgagttaacacag 300
 E  K  D  E  L  T  Q  K  N  H  E  L  T  R  Q  I  N  D
gagaaggatgagttaacacagaaaaatcatgagttaacacgncagattaatgat       354
```

Figure 1. DNA and translated amino acid sequence for M30 isolate, 1GL251. 'n' indicates an unknown nucleotide. The first six nucleotides represent the *PstI* cloning site. The asterisk* indicates the beginning of the mature protein. Repeat regions are underlined. The italicised amino acid sequence was determined from PcpM30, as previously described [31].

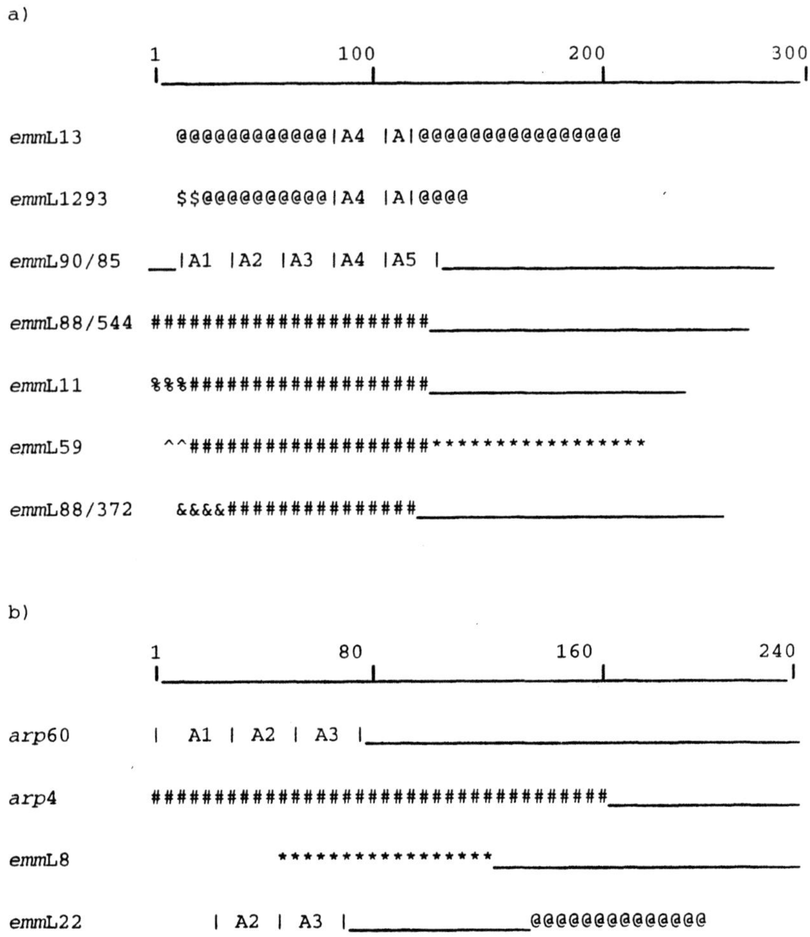

Figure 2. Diagrammatic representation of the region of DNA homology present between the coding sequences of a) the TIN family and b) the ML60 family. The number of bases is indicated. Different symbols are used to show areas of nonhomology within each family. a) *emm*L90/85, *emm*L13 and *emm*L1293 have repetitive sequences that correspond to A repeats of seven amino acids. b) *arp*60 and *emm*L22 have repetitive sequences that correspond to A repeats of seven amino acids.

```
                                                A1       A2

a                                            |______| |______|


b

M5           TVTRGTINDPQRAKEALDKYEHE NHDEKTK NEGLKTE     37
M6          RVFP---VEN-DK-R-L-N--DV- -SMLQAN -DK-T--     38
M30         TII--DIAR--DKLRKIAENF-VQ --QLTQE K---TRK     38
STNS1       RVT--SQAQ-AAGL--KA----VR --ELEHN --K----     38
ML18         PL--A-ADNKDELIKRAND--IQ --QLTVE -KKI--D     37



               A3       A4       A5

a           |______| |______| |______|


b

M5          NEGLKTE NEGLKTE NEGLKTE KKEHEAENDKLKQQRDTLS     78
M6          -NN-TDQ -KN-T-- -KN-TDQ N-NLTT--KN-TD-NKN-T     79
M30         -GE-TQ- K                                       46
STNS1       -SD---- -SK-TS- K-E-TQ- -EDTDQ--R               66
ML18        K-Q-.-K RMMI-L- KDQ-D-                          57
```

Figure 3. Alignment of translated amino acid sequences of the M5 family. a) The A repeats of M5 are seven amino acids in length. b) Amino acid alignment for M5 [9], M6 [7], ML18 [32], M30 and STNS1 [27]. The dashes show common amino acids and the dots indicate gaps in the sequence. The bolded and italicised sequence shows the region of nucleotide homology. The repeats present for each M or ML are underlined. The number of amino acids is shown on the right of the figure.

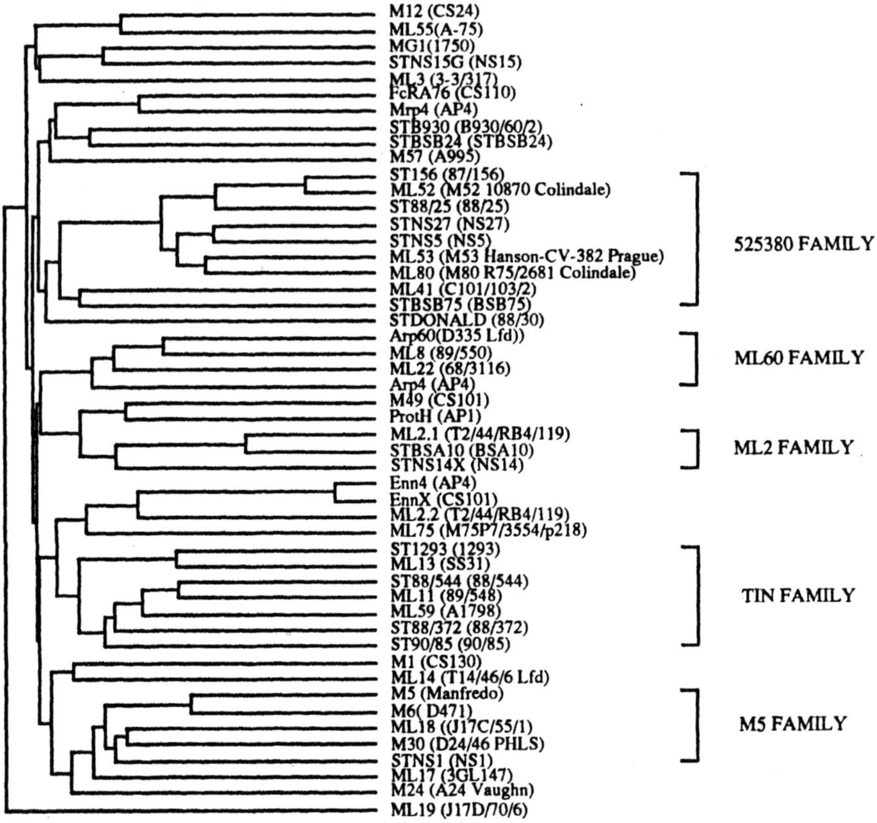

Figure 4. Dendrogram of all M and ML sequences. The dendrogram was constructed using PILEUP [40]. The amino acid sequences were truncated prior to the start of the C repeat region. The isolate from which the sequence was derived is shown in the parentheses, and where available was taken from the published sequences; M1 [8], ML2.1 and ML2.2 [33], ML3 [36], M5 [9], M6 [7], ML8, ML11, STDONALD, STNS1, STNS14X, ST88/25, ST90/85 and ST1293 [27], M12 [6], ML18, ML19, STB930 [32], M24 [10], ML41 [20], M49 and EnnX [11], M57 [31], Arp4 [29], Enn4 [19], FcRA76 [16], Mrp4 [35], MG1 [34], ProtH [18]. The five families which also showed localised homology by an alternate method of analysis, are indicated.

```
a
     1GL251        TIITRAIAKDPDKLKEIAENFEVQNH
     2054          -----D--R-----RK----------

b
     1GL251        A1     ELTRKNGELTQEKD
                   A2     --------------
                   A3     ---HE---------
                   A4     ---QK-H---RQIN
     2054          A'1    .......Q-----E
                   A2     G-------------

c
     88/25         A1     SENEELQAGLQEKE
     90/86         A1     -------------K
                   A2     R-------------
```

Figure 5. Alignment of a) N terminal nonrepetitive regions, and b) A repeat region for M30 isolates 2054 and 1GL251, and c) A repeat regions for ST88/25 isolates 88/25 and 90/86. The dashes show common residues.

2 G.N. Phillips Jr., P.F. Flicker, C. Cohen, B.N. Manjula and V.A. Fischetti; "Streptococcal M protein: alpha-helical coiled-coil structure and arrangement on the cell surface"; *Proceedings of the National Academy of Sciences U.S.A.*; "Vol.78"; 1981; p.4689.

3 R.C. Lancefield; "Persistance of type-specific antibodies in man following infection with group A streptococci"; *Journal of Experimental Medicine*; "Vol.110"; 1959; p.271.

4 A.G. Kuttner and E. Krumwiede; "Observations on the effect of streptococcal upper respiratory infections on rheumatic children: a three year study"; *Journal of Clinical Investigation*; "Vol. 20"; 1941;p. 273.

5 R.C. Lancefield; "Current knowledge of type-specific M antigens of group A streptococci"; *Journal of Experimental Medicine*; "Vol.89"; 1962; p.307.

6 J.C. Robbins, J.G. Spanier, S.J. Jones, W.J. Simpson and P.P. Cleary; "*Streptococcus pyogenes* type 12 M protein gene regulation by upstream sequences"; *Journal of Bacteriology*; "Vol.169"; 1987; p.5633.

7 S.K Hollingshead, V.A. Fischetti and J.R. Scott; "Complete nucleotide sequence of type 6 M protein of the group A streptococcus. Repetitive structure and membrane anchor"; *Journal of Biological Chemistry*; "Vol.261"; 1986; p.1677.

8 E. Haanes-Fritz, W. Kraus, V. Burdett, J.B. Dale, E.H. Beachey and P. Cleary; "Comparison of the leader sequences of four group A streptococcal M protein genes"; *Nucleic Acids Research*; "Vol.16"; 1988; p.4667.

9 L. Miller, L. Gray, E. Beachey and M. Kehoe; "Antigenic variation among group A streptococcal M proteins. Nucleotide sequence of the serotype 5 M protein gene and its relationship with genes encoding types 6 and 24 M proteins"; *Journal of Biological Chemistry*; "Vol.263"; 1988; p.5668.

10 A.R. Mouw, E.H. Beachey and V. Burdett; "Molecular evolution of streptococcal M protein: cloning and nucleotide sequence of the type 24 M protein gene and relation to other genes of *Streptococcus pyogenes"; Journal of Bacteriology*; "Vol. 170"; 1988; p.676.

11 E.J. Haanes and P.P. Cleary. "Identification of a divergent M protein gene and an M protein-related gene family in *Streptococcus pyogenes* serotype 49"; *Journal of Bacteriology*; "Vol.171"; 1989; p.6397.

12 B.N. Manjula, B.L. Trus and V.A. Fischetti; "Presence of two distinct regions in the coiled-coil structure of the streptococcal Pep M5 protein: Relationship to mammalian coiled-coil proteins and implications to its biological properties"; *Proceedings of the National Academy of Sciences U.S.A.*; "Vol.82"; 1985; p.1064.

13 B.N. Manjula, A.S. Acharya, T. Fairwell and V.A. Fischetti; "Antigenic domains of

the streptococal Pep M5 protein. Localization of epitopes crossreactive with type 6 M protein and identification of a hypervariable region of the M molecule"; *Journal of Experimental Medicine*; "Vol.163"; 1986; p.129.

14 E.H. Beachey and J.M. Seyer; "Protective and nonprotective epitopes of chemically synthesized peptides of the NH2-terminal region of type 6 streptococcal M protein"; *Journal of Immunology;* "Vol.136"; 1986;p.2287.

15 J.B. Dale and E.H. Beachey; "Localization of protective epitopes of the amino terminus of type 5 streptococcal M protein"; *Journal of Experimental Medicine*; "Vol.163"; 1986; p.1191.

16 D.G. Heath and P.P. Cleary; "Fc-receptor and M-protein genes of group A streptococci are products of gene duplication"; *Proceedings of the National Academy of Sciiences U. S. A;* "Vol.86"; 1989; p.4741.

17 G. Lindahl; "Cell surface proteins of a group A streptococcus type M4: the IgA receptor and a receptor related to M proteins are coded for by closely linked genes"; *Molecular and General Genetics*; "Vol.216"; 1989; p.372.

18 H. Gomi, T. Hozumi, S. Hattori, C. Tagawa, F. Kishimoto and L. Bjorck; "The gene sequence and some properties of protein H, a novel IgG-binding protein"; *Journal of Immunology*; "Vol.144"; 1990; p.4046.

19 H. Jeppson, E. Frithz and L.O. Heden; "Duplication of a DNA sequence homologous to genes for immunoglobulin receptors and M proteins in *Streptococcus pyogenes*"; *FEMS Microbiological Letters*; "Vol.71"; 1992; p.139.

20 A. Podbielski; "Three different types of organization of the vir regulon in group A streptococci"; *Molecular and General Genetics*; "Vol,237"; 1993;p. 287.

21 J.R. Scott; "The M protein of group A streptococcus: evolution and regulation"; In: *The Bacteria; Volume XI*; B. Iglewski (Ed.); New York: Academic Press, Inc; 1990; pp. 117-203.

22 M.P. Harbaugh, A. Podbielski, S. Hugi and P.P. Cleary; "Nucleotide substitutions and small-scale insertion produce size and antigenic variation in group A streptococcal M1 protein"; *Molecular Microbiology*; "Vol.8"; 1993; p.981.

23 V.A. Fischetti, M. Jarymowycz, K.F. Jones and J.R. Scott; "Streptococcal M protein size mutants occur at high frequency within a single strain"; *Journal of Experimental Medicine*; "Vol.164"; 1986; p.971.

24 S.K. Hollingshead, V.A. Fischetti and J.R. Scott; "Size variation in group A streptococcal M protein is generated by homologous recombination between intragenic repeats"; *Molecular and General Genetics*; "Vol.207"; 1987; p.196.

25 K.F. Jones, S.K. Hollingshead, J.R. Scott and V.A. Fischetti; "Spontaneous M6

protein size mutants of group A streptococci display variation in antigenic and opsonogenic epitopes:; *Proceedings of the National Academy of Sciences U. S. A.*; "Vol.85"; 1988; p.8271.

26 W.A. Relf and K.S. Sriprakash; "Limited repertoire of the C-terminal region of the M protein in *Streptococcus pyogenes*"; *FEMS Microbiological Letters*; "Vol.59"; 1990; p.345.

27 W.A. Relf, D.R. Martin and K.S. Sriprakash; "Identification of sequence types among the M nontypeable group A streptococci"; *Journal of Clinical Microbiology*; "Vol. 30"; 1992; p.3190.

28 J. Garnier, D.J. Osguthorpe and B. Robson; 'Analysis of the accuracy and implications of simple methods for predicting the secondary structure of globular proteins"; *Journal of Molecular Biology*; "Vol.120"; 1978; p.97.

29 E. Frithz, L.O. Heden and G. Lindahl; "Extensive sequence homology between IgA receptor and M proteins in *Streptococcus pyogenes*"; *Molecular Microbiology*; "Vol.3"; 1989; p.1111.

30 G. Lindahl and B. Akerstrom; "Receptor for IgA in group A streptococci: cloning of the gene and characterization of the protein expressed in *Escherichia coli"; Molecular Microbiology*; "Vol.3"; 1989; p.239.

31 B.N. Manjula, K.M. Khandke, T. Fairwell, W.A. Relf and K.S. Sriprakash; "Heptad motifs within the distal subdomain of the coiled-coil rod region of M protein from rheumatic fever and nephritis associated serotypes of group A streptococci are distinct from each other: Nucleotide sequence of the M57 gene and relation of the deduced amino acid sequence to other M proteins"; *Journal of Protein Chemistry;* "Vol.10"; 1991; p.369.

32 A. Podbielski, B. Melzer and R. Lutticken; "Application of the polymerase chain reaction to study the M protein(-like) gene family in beta-hemolytic streptococci"; *Medical Microbiology and Immunology (Berlin)*; "Vol.180"; 1991;p. 213.

33 D.E. Bessen and V.A. Fischetti; "Nucleotide sequences of two adjacent M or M-like protein genes of group A streptococci: different RNA transcript levels and identification of a unique immunoglobulin A-binding protein"; *Infection and Immunity*; "Vol.60"; 1992; p.124.

34 C.M. Collins, A. Kimura and A.L. Bisno; "Group G streptococcal M protein exhibits structural features analogous to those of class I M protein of group A streptococci"; *Infection and Immunity*; "Vol.60"; 1992; p.3689.

35 P. O'Toole, L. Stenberg, M. Rissler and G. Lindahl; "Two major classes in the M protein family in group A streptococci"; *Proceedings of the National Academy of Sciences U.S.A.*; "Vol.89"; 1992;p. 8661.

36 A. Podbielski, R. Baird and A. Kaufhold; "The group A streptococcal M-type 3 protein gene exhibits a C terminus typical for class I M proteins"; *Medical Microbiology and Immunology (Berlin)*; "Vol.181"; 1992; p.209.

37 L. Stenberg, P. O'Toole and G. Lindahl; Many group A streptococcal strains express two different immunoglobulin-binding proteins, encoded by closely linked genes: characterization of the proteins expressed by four strains of different M-type"; *Molecular Microbiology*; "Vol.6"; 1992; p.1185.

38 W.J. Wilbur and D.J. Lipman; "Rapid similarity searches of nucleic acid and protein data banks"; *Proceedings of the National Academy of Sciences U.S.A*;,"Vol. 80"; 1983; p.726.

39 S.B. Needleman and C.D. Wunsch; "A general method applicable to the search for similarities in the amino acid sequence of two proteins"; *Journal of Molecular Biology*; "Vol.48"; 1970; p.443.

40 D.F. Feng and R.F. Doolittle; "Progressive sequence alignment as a prerequisite to correct phylogenetic trees"; *Journal of Molecular Evolution*; "Vol.35"; 1987;p. 351.

41 W.A. Relf, D.R. Martin and K.S. Sriprakash; "Antigenic diversity within a family of M proteins from group A streptococci: Evidence for the role of frameshift and compensatory mutations"; *Gene*; submitted

42 G.G. Wiley and P.N. Bruno; "Cross-reactons among group A streptococci. I. Precipitin and bactericidal cross-reactions among types 33, 41, 43, 52 and Ross"; *Journal of Experimental Medicine*; "Vol.128"; 1968; p.959.

43 G.G. Wiley and P.N. Bruno; "Cross-reactions among group A streptococci. II. Further analysis of antigens related to type-specificity and protection"; *Journal of Immunology;* "Vol.103"; 1969;p.149.

44 V.A. Fischetti: "Streptococcal M protein extracted by nonionic detergent. II. Analysis of the antibody response to the multiple antigenic determinants of the M-protein molecule"; *Journal of Experimental Medicine*; "Vol.146"; 1977; p.1108.

45 J.B. Dale and E.H. Beachey; "Unique and common protective epitopes among different serotypes of group A streptococcal M proteins defined with hybridoma antibodies"; *Infection and Immunity*; "Vol.46"; 1984; p.267.

46 K.F. Jones, B.N. Manjula, K.H. Johnston, S.K Hollingshead, J.R. Scott and V.A. Fischett,"Location of variable and conserved epitopes among the multiple serotypes of streptococcal M protein"; *Journal of Experimental Medicine*; "Vol.161"; 1985; p.623.

47 L. Miller, V. Burdett, T.P. Poirier, L.D. Gray, E.H. Beachey and M.A. Kehoe; "Conservation of protective and nonprotective epitopes in M proteins of group A streptococci"; *Infection and Immunity*; "Vol.56"; 1988; p.2198.

RHEUMATIC FEVER ASSOCIATED B-CELL ALLOANTIGEN IN 1854 SCHOOL-AGED HEALTHY CHILDREN

R.S. Rodrigues[1], P. Ontiveros[1], S. Torres[1], and J.B. Zabriskie[2]

ABSTRACT

In previous studies several authors have shown that 90-100% of Rheumatic Fever (RF) patients are positive for a rheumatic fever associated B-cell alloantigen identified by a monoclonal antibody denominated D8/17. In the present work we studied 1854 school aged healthy children to determine the percentage positive for the marker and if this antigen was predictive for the development of RF. After written consent was obtained 0.15 ml of whole blood was drawn into a heparinized tube. A monoclonal antibody (D8/17) specific for the alloantigen was added. We then employed a technique of double staining using first isothiocyanate conjugated goat anti-mouse IgM and then adding, to identify B cells, anti HLA-DR conjugated with phycoerythrin. After lysis of the erythrocytes the results were read under the microscope counting 200 cells. The percentage of cells were the number of positive D8/17 cells that were also B-cells. A test was considered positive when the number of cells with the marker was 2 SD above the average (20% or greater). Eighty seven children (4.7%) were found positive for the marker. None have developed in the first months of follow up RF.

1. INTRODUCTION

During the past decade, there has been a decrease of rheumatic fever in developing countries [1] . However in the United States and after a long period of near absence, the disease has reappeared as shown by several outbreaks in the last few years [2] Of interest in these outbreaks was that pharyngitis was not documented in more than fifty percent of the patients studied [3]. Under these circumstances primary prevention of rheumatic fever could not possibly have been accomplished [4]. Moreover forty percent of the cases where pharyngitis was documented were not properly treated [5]. Other interesting finding in these outbreaks was the reappearance of streptococci type 18 that was related in the past with rheumatic fever outbreaks in military populations

1 Hospital Infantil de México Federico Gómez, calle Dr. Márquez 162 México, D.F. 06720
2 The Rockefeller University, 1230 York Ave. NY NY 10021-6399

[6]. This streptococcal strain has been isolated recently in Mexico City [7]. The emergence of streptococcal strains that exhibit biologic characteristics known to be rheumatogenic and its association with rheumatic fever outbreaks supports the evidence of the rheumatogenicity of certain streptococcal strains. On the other hand host factors play a role in the susceptibility to rheumatic fever [8]. In this context Patarroyo et al [9] identified a B lymphocyte marker in rheumatic fever patients. Zabriskie et al [10] developed monoclonal antibodies in mice that identified the marker in 92% of rheumatic fever patients and in 21% of controls. Later on Khanna et al [11] developed a new monoclonal antibody (D8/17) that identified the B cell marker denominated also alloantigen D8/17 in 100% of rheumatic fever patients and in just 5-7% of the controls. These studies were confirmed in Mexico by Rodriguez et al [12] using the same monoclonal antibody.The objective of this study was to determine the prevalence of alloantigen D8/17 in healthy school aged children and its predictive value for the development of rheumatic fever.

2. MATERIAL AND METHODS.

A population of school children living in an area known to have high rates of rheumatic fever was chosen for the study. A pediatrician gave informative talks about the nature of the project to the parents and school principals of the schools included in the study. After written informed consent was obtained from the parents and schools authorities, blood was drawn by the "finger prick" method.

a) Sample Collection:

Under sterile conditions, 0.15 ml of blood was obtained from each child by a finger stick; the blood was placed into a heparinized capillary tube (Sarsted) and gently mixed. Ten blood samples were drawn every day four times a week during a school calendar year.

b) Immunological Assays:

Each sample was immediately transported to the laboratory and processed within two hours of collection time. To each blood sample we added 100 μl of a 1:20 dilution of monoclonal antibody D8/17 prepared according to methods previously described [11]. After an incubation time of one hour at 4° C the cells were washed and 3 ml of PBS-BSA added to the mixture and the sample centrifuged at 1200 rpm for three minutes. The supernatant was discarded and the cell pellet incubated with 20 μl of a 1:5 dilution of

FITC-antimouse IgM (Sigma Imunochemicals, St Louis Mo, USA). After an incubation time of 45 minutes at 4º C cells were washed again as previously described then, 20 μl of a 1:5 dilution of phycoerythrin conjugated anti-HLA-DR (Anti-human HLA DR-PE, Beckton Dickinson Immunocytometry Systems, San Jose CA, USA) were added and incubated for 30 minutes at 4º C. Cells were washed again. To achieve erythrocyte lysis ammonium chloride was then added and let react for 4 minutes at room temperature. The cell pellet was washed three times at 1300 rpm, three minutes each time. Five ml of a 2% formaldehyde PBS-solution was then added to the cell pellet and refrigerated at 4º C until its reading under the microscope. For the final reading, 15 μl of the prepared solution was suspended on a glass slide and observed under the fluorescence microscope. Red fluorescent cells (representing B lymphocytes) were counted; the filter was changed and green inmunofluorescence was detected if the D8/17 marker was present in B lymphocytes. A ratio of positive (green) cells /200 B lymphocytes (red fluorescent cells) was calculated for each sample.

A sample was considered positive for alloantigen D8/17 when the ratio so obtained was greater than 2 SD above the mean ratio of 300 samples obtained from healthy children. In our study the presence of 20% or greater cells positive for the D8/17 marker was considered a positive test. Positive and negative controls were included.

c) Follow up:

The parents of children included in the study were informed of the result of the test with an explanation given about its possible significance. Every child with a positive test had a complete physical examination to rule out pre-existing rheumatic fever. All children (positive or negative for the marker) were given a green card to keep. This green card would assure priority care at Hospital Infantil de Mexico for any child who developed signs or symptoms of acute rheumatic fever during the observation period. Every child positive for the marker was scheduled, in addition, to be examined twice a year even in the absence of evidence of rheumatic fever. For each positive patient a negative one was followed for the same purpose to serve as control. The total observation period will be of two years. However the data is being analized every 6 months, the end point being the frequency of rheumatic fever cases in the group of students positive or negative for the marker. To test the significance of the differences in the number of rheumatic fever cases between the two groups, non-parametric statistic tests will be used and the predictive value of a positive or negative test assessed. Here we report the first year of follow up and the prevalence of the alloantigen in this group of patients.

3. RESULTS

From june 1991 to may 1993 we studied 1854 children for the presence of alloantigen D8/17 (TABLE 1). Fifty four percent were males and forty six percent females. The average age of the children studied was 9.6± 1.6 years old. According to the criteria previously outlined 87 children were positive for the marker (4.7%). As of august 1993 and after an average period of follow up of 13 months for the entire group, no child positive or negative for the marker has developed rheumatic fever.

TABLE 1

Cumulative Number of Children Studied According to Different Study Periods

PERIOD	No. CHILDREN STUDIED [Cumulative]	No. POSITIVE [Cumulative]
February 1991-May 1991	464	29
September 1991-November 1991	773	40
January 1992-May 1992	1419	59
January 1993-May 1993	1854	87 (4.7%)

4. DISCUSSION

In the past it has been shown that the monoclonal antibody labelled D8/17 appears to identify a cell-surface marker which shows markedly increased expression on the B-cells of all individuals previously diagnosed as having had rheumatic fever or rheumatic heart disease [10]. The cell surface marker recognized thus far appears to be unique to B-lymphocytes since it was not shown to be present on T-cells, macrophages or non-B cell populations isolated from rheumatic fever patients [11].

The finding of 4.7% of healthy children positive for the marker was obtained using as a criteria for a positive patient a number of B-lymphocytes with the alloantigen D8/17 2 S.D above the average. However we will know if this criteria was correct when we analyze at what percentage of B-cells positive for the marker the rheumatic fever cases appeared. The predictive value of a positive or negative test remains to be defined until a larger number of patients are included and a longer follow up period is assured. However if the patients positive for the marker are shown to develop acute rheumatic fever in contrast to the negative ones then the predictive value of a positive test would have been established. In this context a study by Regelmann et al [13] of rheumatic fever patients and their families revealed that one unaffected individual in a rheumatic family who was markedly positive for the D8/17 marker went on to contract the disease. The early identification of the susceptible population coupled with careful monitoring of the incidence of streptococcal infection could lead to a more effective antibiotic strategies thus enhancing the prevention of the disease. However, as we have noted previously, the number of children with a clinical history of pharyngitis preceding acute rheumatic fever in recent outbreaks was less than fifty percent, making primary prevention a difficult task to accomplish. Bisno [2] has stated that given the devastating effects of acute rheumatic fever in the Third World and the resurgence of invasive streptococcal disease and rheumatic fever in the United States, the development of a multivalent M protein vaccine [14,15] against the most virulent and rheumatogenic serotypes seems a valid public health objective. In order for a streptococcal vaccine to be accepted it would have to be devoided of cross reactions with mammalian muscle proteins. Because these muscle proteins, have been shown in some animals to have common conformational features with the M protein of the streptococci, it is likely that a cross reactive antibody could recognize both structures simultaneously. On the other hand the study of the molecular structure of alloantigen D8/17 by Zabriskie et al [16] has shown that this protein has 701 aminoacids and a unique helical coil-coiled structure with a high conformational homology with both cardiac and cytoskeletal and streptococcal M protein moieties. These recent findings suggests that an M protein vaccine should be tried only in children negative for the marker in order to avoid potential cross reactions with the proteins of the host that could lead theoretically to damage of the heart and other host tissues.

5. CONCLUSIONS

The prevalence of the D8/17 B-lymphocyte marker in 1854 healthy children studied in Mexico City was 4.7 percent. The predictive value of a positive test has not yet been established. A high conformational homology between the D8/17 marker, M protein and cytoskeletal proteins has been demostrated which precludes the use of an M protein vaccine in children positive for this alloantigen.

ACKNOWLEDGEMENT

This work was supported in part by the Thrasher Foundation, Salt Lake City, Utah, USA

REFERENCES

1. M. Markowitz " Streptococcal disease in developing countries" *Pediatr Infec Dis J.* Vol. 10, Nº 10 [Suppl] 1991; p. S11-4.

2. A.L. Bisno. "Group A streptococcal infections and acute rheumatic fever". *N. Engl J Med.* Vol. 325, Nº 11, 1991; p. 783-93.

3. L.G. Veasy, S.E. Wiedmeir, G.S. Orsmond, H.D. Ruttenberg, M.M. Boucek, S.J. Roth, V.F. Tait, J.A. Thompson, J.A. Daly E.L. Kaplan et al. "Resurgence of acute rheumatic fever in the intermountain area of the United States". *N. Engl J Med.* Vol. 316, Nº 8, 1987; p. 421-7.

4. A.S. Dajani. "Current Status of Nonsuppurative complications of Group A Streptococci" *Pediatr Infect Dis J.* Vol 10, Nº 10, [Suppl]: 1991; p. 525-7.

5. L.G. Veasy and H.R. Hill. " Clinical Features of acute rheumatic fever recrudescence in the intermountain area of the United States". *Presented at the XI Lancefield Symposium on Streptococci and Streptococcal Diseases.* Siena, Italy, September 10-14, 1990.

6. L. James and R.B. Mc Farland. "An epidemic of pharyngitis due to nonhemolytic group A streptococcus at Lowry air force base". *N Engl J Med.* Vol 284, Nº 14, 1971; p. 750-2.

7. R.S. Rodríguez, P. Ontiveros, S. Torres, A.K. Khanna, D.R. Buskirk, and J.B. Zabriskie. "Presencia de un antígeno no HLA en los linfocitos B de pacientes con fiebre reumática y sus familiares definido mediante anticuerpos monoclonales" *Bol Med Hosp Infant Mex.* Vol 47, Nº 5, 1990; p. 313-7.

8. E.M. Ayoub, D.J. Barret, N.K. Maclaren, and J.P. Krischer "Association of class II histocompatibility leukocyte antigens with rheumatic fever" *J Clin Invest.* Vol 77, Nº 6, 1986; p. 2019-26.

9. M.E. Patarroyo, R.J. Winchester, A. Vejerano, A. Gibofsky, F. Chalem, J.B. Zabriskie and H.G. Kunkel. "Association of a B-cell alloantigen with susceptibility to rheumatic fever". *Nature*. Vol 278, Nº 5700; 1979, p. 173-4.

10. J.B. Zabriskie, D. Lavenchy, R.C. Williams Jr, S. M. Fu, CA Yeadon, M. Fotino, and D.G. Braun. "Rheumatic fever-associated B cell alloantigens as identified by monoclonal antibodies" *Arthritis Rheum*. Vol 28, Nº 9, 1985; p. 1047-51.

11. A.K. Khanna, D.R. Buskirk, A. Gibosky, A. Menon. M. Fotino H.M. Reed et al. "The presence of a non-HLA B-cell antigen in rheumatic fever patients and their families as defined by a monoclonal antibody". *J Clin Invest*. Vol 83, Nº 5, 1989; p. 1710-6.

12. R.S. Rodríguez, J.E. Mendez-Cisneros, L.E. Espinosa, M. Pecero, and A-Bahamon. "Utilidad de la penicilina benzatínica combinada en el tratamiento de la faringoamigdalitis estreptocócica". *Bol Med Hosp Infant Mex*. Vol 45, Nº 12, 1988; p. 797-802.

13. W.E. Regelmann, E.D. Gray, R. Talbot, L. Cairns, D. Martin, L.C. Miller, J.B. Zabriskie et al. "Distribution and inheritance of rheumatic antigens in peripheral blood of RHD patients and their families". *Xth Lancefield International Symposium on Streptococci and Streptococcal Diseases* , 1987. Abstract 58, page 85.

14. J.B. Dale and E.H. Beachy. "Epitopes of streptococcal M proteins shared with cardiac myosin". *J Exp Med*. Vol 162, Nº 2, 1985, p. 583-591.

15. V.A. Fischetti. "Streptococcal M protein". *Scientific Amer*, Vol. 264, Nº 6, 1991, p. 58-65.

16. J.B. Zabriskie. Personal communication, august, 1993.

HETEROGENEITY OF GROUP A STREPTOCOCCAL MI CHROMOSOMAL RELP PATTERNS AS DETERMINED BY PEGE

A.N. SUVOROV AND J.J. FERETTI

[1]Institute of Experimental Medicine, St. Petersburg, Russia
[2]University of Oklahoma Health Sciences Center, Oklahoma City, OK USA

INTRODUCTION.

Group A streptococci (GAS) continue to cause serious and sometimes fatal infections in children and adults, even after invention with penicillin [2]. Recently, there has been a striking increase in the mortality rate caused by streptococcal infections [1,4,5]. In spite of the clinical significance of GAS infections, precise diagnostic procedures for identifying and characterizing streptococcal strains is presently performed in only a few highly specialized centers. The most commonly accepted method of characterizing streptococcal strains is by serological identification of M protein, which requires a comprehensive collection of sera (more than 100). The complexity of strain characterization is increasing since recent reports have shown that strains belonging to the same M serotype may possess an M protein which differs in size [3] and genetic structure [4,5].

Cleary and coworkers [1] analyzed genomic DNA from 67 GAS strains belonging to the same serotype (M1) by conventional restriction endonuclease analysis. Their results suggested that strains associated with serious disease (invasive) originated from a unique clone and could be distinguished from strains associated with uncomplicated disease (non-invasive variants).

Recently, Martin and Single [9] used restriction fragment length poylmorphism (RFLP) profile analysis of genomic DNA to show that at least six different RFLP patterns were present among 90 different M1 GAS strains isolated in New Zealand. One RFLP pattern accounted for 74% of the strains analyzed. In this study, we have employed RFLP anlaysis of genomic DNA of M1 GAS strains from nine different countries using pulsed field gel electrophoresis (PFGE).

MATERIALS AND METHODS

Bacteria. Streptococcal M1 strains tested were obtained from several collections representing nine different countries; including J. J. Ferretti, the University of Oklahoma (20 strains); J.Zabriskie, Rockefeller University, (10 strains); I. Iontova, Institute Experimental Medicine, St.Petersburg, Russia (7 strains); C. Schaleen, Lund Institute of Medical Microbiology, Sweden, (3 strains); S.Scriver, Mount Sinai Hospital, University of Toronto, Canada (11 strains); P.P.Cleary, Department of Microbiology, University of Minnesota (4 strains).

All the M1 strains tested were identified by serological M or T typing and the M type was confirmed using the polymerase chain reaction (PCR) [8] employing oligonucleotide primers specific for *emm1*. Streptococcal strains were cultivated in Todd-Hewitt broth at 37 C; in several cases 5% horse serum was added for better growth.

Pulsed field electrophoresis (PFGE). All M 1 strains used in the experiments were subjected to PFGE analysis. Chromosomal DNA preparation was carried out using a previously developed protocol [7]. DNA embedded into 2 mm3 agarose plaques was digested overnight at room temperature using restriction endonuclease *SmaI* (20 u/mcg). The whole procedure, beginning with cell inoculation and finishing with PFGE analysis, usually took about a week. However, once DNA was embedded in the agarose it could be stored in 200mM EDTA for several months without degradation.

DNA fragments were separated by PFGE employing either a BIO RAD or Beckman Gene Line electrophoresis system. The parameters for operation were the same for both systems; i.e., 200 V with ramp from 10 to 40 sec for 22 hr. A lambda DNA ladder was used as a molecular weight marker, using 5 mcg of phage DNA in solution (0.5mcg/mcl, BRL). Molecular weight estimations were done using regression analysis by the Seqaid computer program.

Restriction enzymes were purchased from Boehringer Mannheim (FRG) or Biopol (Russia). PCR amplification of *emm* was carried out using *Taq* polymerase from Promega (USA) or *Tth*I from Dia-M (Russia). Proteinase K and lyzosyme were obtained from Sigma (USA). Following electrophoresis and photographic recording, DNA fragments from the gel were transferred to nylon membranes by the procedure of Southern [6] and hybridized with *emm*1 or *scp*B gene probes, labeled using the Boehringer Mannheim Genius kit for nonradioactive labeling with digoxigenin. PCR was carried out using a Perkin-Elmers or Techne thermal cycler.

RESULTS AND DISCUSSION

The restriction enzyme, *Sma*I, was used to digest genomic DNA in these studies since it resulted in only 11-12 fragments, ranging in size from 50 to 500 kb. A PFGE separation of the *Sma*I digested DNA fragments from one of the M1 GAS strains is presented Fig.1. M1 strains analyzed in this study were obtained from nine countries isolated during several different time periods (Table 1). Most of the strains isolated in 1991-92 were from patients known to have serious or invasive infections, while strains isolated prior to that time represented a cross-section of streptococcal diseases.

Table 1. M1 GAS strains.

Strain Number	Country of Origin	Year Isolated
1 - 20	7 countries*	1980-85
21 - 30	USA	1933-73
31 - 37	Russia	1992
38 - 40	Sweden	1992
41 - 50	Canada	1991
51 - 54	USA**	1991-92

. * Canada, Denmark, England, Germany, New Zealand, Thailand, USA
** Strains 51-52 were invasive; 53-54 were non-invasive

In order to organize the PFGE data and compare different restriction length polymorphism (RFLP) patterns, we have chosen a way of representing these patterns in a soccer-like formation; e.g., 4-3-3, 4-2-4, etc. According to this scheme, restriction fragments that were grouped together were given a number representing the number of fragments in that group, the next group assigned a number of the fragments together, etc. Thus, the pattern of a typical RFLP might be represented as 4(210, 200, 190, 170 kb) - 2 (120, 100, 95 kb) - 3 (60, 50, 30 kb). The overall results of RFLP patterns of M1 strains analyzed in this study are presented in Table 2.

Table 2 . RFLP patterns of the analyzed strains

M1 strain No	Pattern No.	RFLP	Sizes (kb)	*emm*1 location
1,3-5,7,23,43, 44,47,50-52	1	4-2-3	200-100-50	200
2	2	4-2-1-4	200-120-130-50	130
9	3	5-1-5	250-130-50	130
10-13,26,41,42, 45,46,48,49,53-55	4	4-1-4	200-130-50	180
14,-17,24*,25	5	1-2-2-3	220-170-130-50	130
18,27*,	6	1-5-4	250-140-50	180
29,31-40	7	3-2-4	200-100-50	180

* strains with 1 restriction fragment difference in the pattern, but similar emm fragment location

Preliminary RFLP results demonstrated a high degree of heterogeneity among the M1 c strains. However, after analysis of the strains from the OU collection we were able to distinguish at least 7 different RFLP patterns. In order to confirm that all the strains were in fact M1 strains, we hybridized *SmaI* restriction fragments from the analyzed strains with digoxigenin labeled emm1 gene probe. Most of the strains contained emm1 gene on the 200,180, or 130 kb fragments (Table 2).

The most common RFLPs found among these strains were patterns 1,4,5, and 7 . The strains obtained from Russia (nn31-37) and Sweden (nn38-40) were clustered in pattern 7, and the strains from Canada were found in patterns 1 and 4; each group representing a more or less unified group which came from the same region of the country at the same time. Four strains obtained from P.Cleary represented "invasive" strains (51,52) and "non-invasive" (53,54) strains and were in RFLP patterns 1 and 4 respectively. The location of the emm gene was on a different fragment in the two patterns and these results are in agreement with those of Cleary et al. (1) who found similar difference among two serotypes

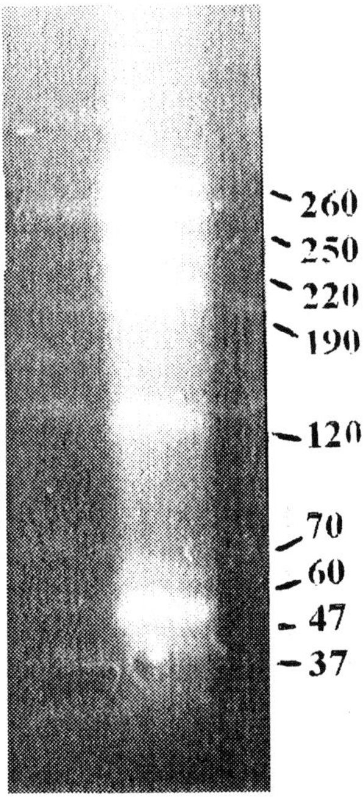

Fig.1.
M1 streptococcal DNA (strain n.12), digested with SmaI

employing conventional electrophoresis. However, we found no correlation of RFLP pattern and disease association in the overall analysis of strains. The only exclusion were the six strains in this study associated with acute rheumatic fever, which seems to possess significant genomic differences. All of these findings suggest that there is considerable genomic heterogeneity throughout the M1 serotype. On the other hand, streptococcal strains collected from one geographical zone during a limited period of time turned out to be quite similar in RFLP pattern. For example, All the strains from the Russian and Swedish collection belonged to the same RFLP pattern. This finding suggests that PFGE analysis of genomic DNA of strains may be of particular importance in future epidemic surveillance and epidemiological studies. Finally, PFGE RFLP patterns are easy to read and the protocol of making streptococcal genomic DNA preparations is simple and reproducible.

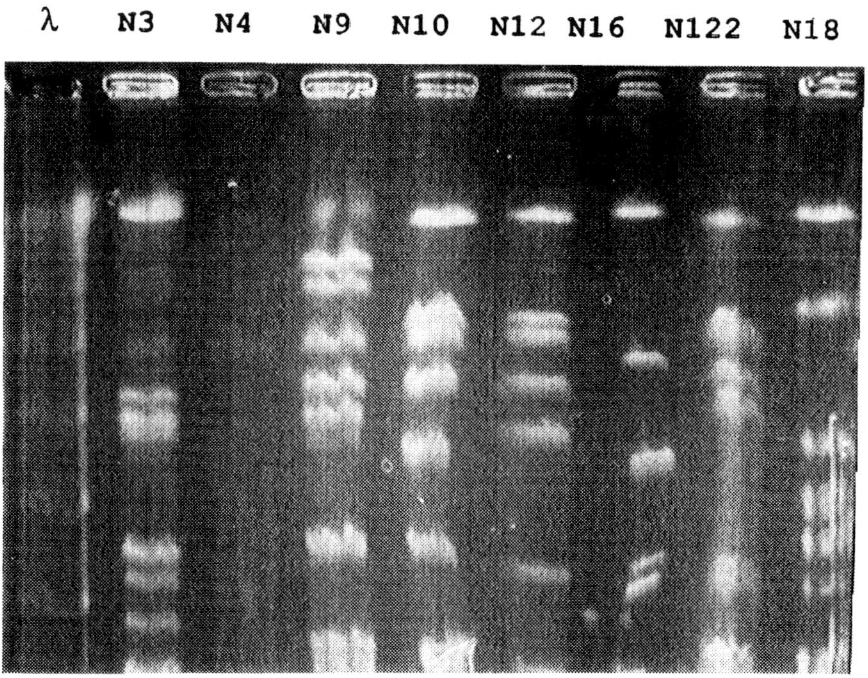

Fig.2.

PFGE analysis of different M1 strains DNAs digested with SmaI

REFERENCES

1. Cleary P.P., Kaplan E.L., Handley J.P., Wlazlo A., Kim M.H., Hauser A.R., and Schlievert P.M.. 1992. Clonal basis for resurgence of serious Streptococcus pyogenes disease in the 1980s. Lancet 339:518.

2. Holm E.S., Norrby A., Bergholm A-M., and Norgen M.. 1992. Aspects of pathogenesis of serious group A streptococcal infection in Sweden,1988-1989. JID. 166:31.

3. Jones K.F., Hollingshead S.K., Scott J.R., and Fischetti V.A.. 1988. Spontaneous M6 protein size mutants of group A streptococci display variation in antigenic and opsonogenic epitopes. Proc. Natl.Acad. Sci. USA 85:8271.

4. Nogren M., Norrby A., and Holm S.E.. 1992. Genetic diversity in T1M1 group A Streptococci in relation to clinical outcome of infection. JID 166:1014.

5. Single L.A., and Martin D.R.. 1992. Clonal differences within M-types of the group A Streptococcus revealed by pulsed field gel electrophoresis. FEMS Microbiol. Lett. 91:85.

6. Southern E.M. 1975. Detection of specific sequences among DNA fragments separated by gel electrophoresis. J. Mol. Biol. 98:503.

7. Suvorov A.N., and Ferretti J.J.. 1993. Genetic analysis of Streptococcus pyogenes M49 chromosomal DNA. Genetics and Molecular Biology of Streptococci, Lactococci, and Enterococci 233.

8. Suvorov A.N., Sverdlova A., and Ferretti J.J.. 1993. A new method for M1 typing using PCR. in preparation

Rapid Identification of Group A Streptococci by PCR

A.N. Suvorov, A.M. Sverdiova, A.A Totolian and J.J. Feretti

SUMMARY

A rapid procedure for the identification of group A streptococci has been developed employing the polymerase chain reaction (PCR). The novel aspect of this procedure is that cells are subjected to the heat treatment step prior to actual PCR, which obviates a DNA isolation steps from bacterial cells. Utilizing this method, several genes were amplified from a variety of strains. This technique should be useful for PCR amplification of streptococci as a routine microbiological procedure for M typing of epidemiological important strains.

INTRODUCTION

The polymerase chain (PCR) reaction is a powerful technique that has been used for numerous procedures involving analysis of individual genes. In the streptococci, PCR has been used for cloning (*scpB*) (12), gene analysis (*slo, speA, speB, scpB*)(1-3,13), sequencing of previously cloned genes (*ska, speA, speB*(2,6,10)), and identification of M type (8). Purified DNA has been used for most of these procedures, and its use requires additional reagents and time consuming steps. For routine analysis of a large number of strains, such as in M typing or identification of the presence of certain genes, a more simplified procedure of access to DNA for PCR amplification is desirable. Recently, a rapid procedure for PCR amplification of streptococcal DNA has been described (7) which employs DNA isolation and several simple purification steps. In this report, we describe a simple heat treatment step of streptococcal cells which allows direct PCR amplification without a DNA purification step.

MATERIALS AND METHODS

44 group A streptococcal (GAS) strains belonging to different serotypes have been tested. These strains were obtained from the collections of the Department of Microbiology and Immunology University of Oklahoma , USA and Department of Molecular Microbiology, Institute of Experimental Medicine, St.Petersburg, Russia.
Bacteria were grown in Todd-Hewitt broth at 37°C overnight. Streptococcal cells from 0.15 ml of culture were precipitated in the microfuge and suspended in 0.1 ml of S buffer (0.5%Triton X-100 in dH_2O). The cell suspension was heated at 93°C for 5 minutes and 5 mcl of it was added to 45 mcl of PCR mixture containing the PCR buffer, with 0.8mMdNTP and 2 units of Taq or Tth polymerase. The enzymes were purchased from Promega (USA) or Bion (Russia).
Primers were syntetised in the Microchemical facilities University of Oklahoma.
Primers for plasmin receptor gene were 5' GGTGGATTTGAAGTAAACGG -forward primer and 5' AACTCAGTTACTGATCCAG - reverse primer.
Primers for *emm*1 amplification were: 5'GGTGATGGTAATCCTAGGGAAGTTATAGAA -

forward primer and 2 reverse primers - 5'TGTTTCACCTGTTGATGGTAACTGTCT and 5'TCTAGTAATCGTTTCTAACTCTTTTCT

RESULTS AND DISCUSSION

In a clinical laboratory setting, there is a need for rapid and simple procedures that can be used with newly evolving techniques, such as PCR. One of our main objectives is to develop alternative procedures for the identification and serotyping of group A streptococci, specifically using PCR. Towards this end, it is desirable to have an efficient one-step method of lysing or teating cells such that DNA is available for amplification. In this study, we have investigated several different lysis methods which might be employed in such a procedure with whole cells; including high temperature, alkaline, and detergent treatments. In addition, two different types of DNA polymerase were employed with the above procedures.

The procedure that was shown to provide the most consistent results utilized S buffer as the lysing solution with a pretreatment of cells at 93°C (for details see materials and methods). Employing the primers for the plasmin receptor gene (*plr*) (9), amplification of an expected 530 bp fragment was accomplished with all GAS strains tested (Fig .1).

Amplification of the *emm* gene was also accomplished using this procedure, employing two set of primers; one with an expected size of 500 bp and other with an expected size of 1300 bp.

The forward primers were the same as those described by Podbielski et al. (1990) (11) and the reverse primers were based on the sequence determined by Haanes-Fritz et al. (1988) (4) and Hollingshead et al.(5). In each case, amplification of fragments of the appropriate size were obtained from a number of different strains (Fig.2). This approach can be valuable for the purpose of both identification of the epidemiological related strains and for the purpose of the fast M typing of the GAS strains.

For epidemiological analysis together with classical PCR which requires 2 primers we used one primer PCR. This approach produces a number of PCR fragments which can be used for the strain finger printing similar to restriction fragments length polymorphism analysis. One primer PCR using our rapid amplification procedure was working equally well (data not shown). The same protocol nicely worked utilizing other microbial species but streptococcus. We have an experience of amplifying the DNA fragments from *Salmonella tiphimurium.*

The method of PCR with S buffer is so simple and reliable that it can be introduced into the clinical laboratories which can be of use for microbiological and epidemiological analysis of pathogenic streptococci.

Fig.1
Rapid amplification of the fragment of plasmin receptor gene from group A streptococcus,
Lane 2-9 different group A streptococus isolates
Lane - λ DNA digested with HindIII, lane 10 - λ DNA digested with StyI

1 2 3 4 5 6 7 8 9 10 [ ]

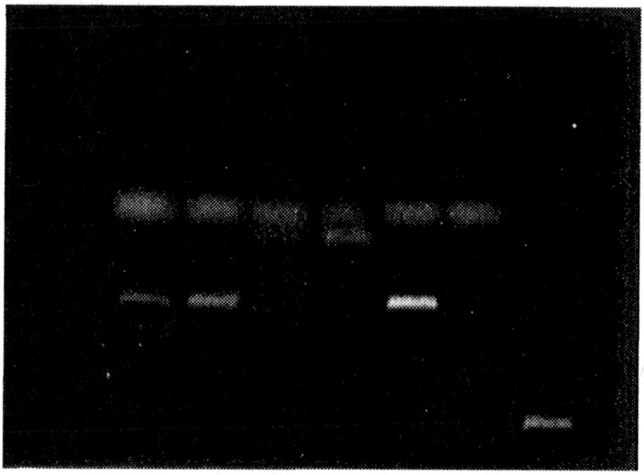

Fig.2
Rapid PCR M typing using emm1 specific primers (amplification of the 500 bp fragment)

Lane 1 λ DNA digesterd with HindIII+EcoRI
Lanes 2,3,6,7 - GAS strains from M1 serotype
Lanes 4,5 - GAS strains from M49 serotype

1 2 3 4 5 6 7

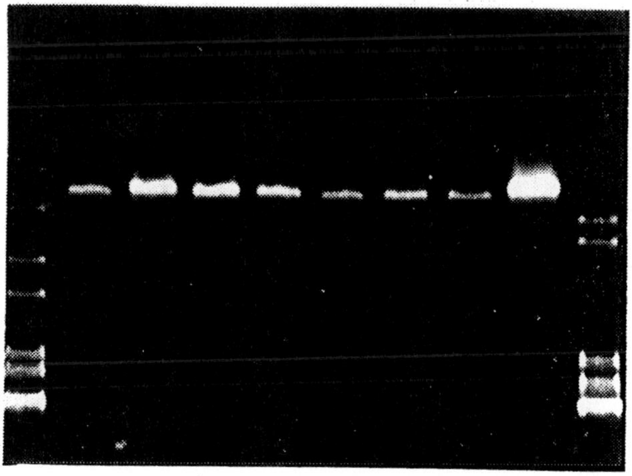

REFERENCES

1. Chang-En Yu and Ferretti J.J. 1989.Molecular epidemiologic analysis of the type A streptococcal exotoxin (erythrogenic toxin) gene (speA) in clinical Streptococcus pyogenes strains. Infect.Immun. 57:3715-3719,

2. Chang-En Yu and Ferretti J.J. 1991.Frequency of the erythrogenic toxin B and C genes (speB and speC) among clinical isolates of group A streptococci. Infect.Immun. 59:211-215,

3. Cleary P.P., Handley J., Suvorov A.N., Podbielski A. and Ferrieri P. 1992.Similarity between the Group B and A streptococcal C5a Peptidase genes. Infect.Immun. 60:4239-4244,

4. Haanes-Fritz, E.W., Kraus, Burdett, Dale J.B., Beachey E.H. and Cleary P.P. 1988.Comparison of the leader sequences of four group A streptococcal M protein genes. Nucleic Acids Res. 16:4667-4677,

5. Hollingshead S.K., Fischetti V.A. and Scott J.R. 1986.Complete nucleotide sequence of type 6M protein of the group A streptococcus: repetitive structure and membrane anchor. J.Biol.Chem 261:1677-1686,

6. HuangT.T., Malke H. and Ferretti J.J. 1989.Heterogeneity of the streptokinase gene in group A streptococci. Infect.Immun. 57:502-506,

7. Hynes W.L., Ferretti J.J., Gilmore M.S. and Segarra R.A. 1992.PCR amplification of streptococcal DNA using crude cell lysates. FEMS Microbiol.Lett. 94:139-142,

8. Kaufhold A., Podbielski A., Johnson D., Kaplan E.L. and Luttiken R. 1992.M protein gene typing of streptococcus pyogenes by nonradioactively labeled oligonucleotide probes. J.Clin.Microbiol. 30:2391-2397,

9. Lottenberg R., Broder C.C., Boyle M.D., Kain S.J., Schroeder B.L. and CurtissIII R. 1992.Cloning,sequence analysis, and expression in Escherichia coli of Streptococcal plasmin receptor. J.Bacteriol. 174:5204-5210,

10. Malke H., Roe B. and Ferretti J.J. 1985.Nucleotide sequence of the streptokinase gene from Streptococcus equisimilis H46A. Gene 34:357-362,

11. Podbielski A., Kuhnemund O. and Lutticken R. 1990.Identification of group A type 1 streptococcal M protein gene by a non-radioactive oligonucleotide detection method. Med.Microbiol Immunol 179:255-262,

12. Suvorov A.N., Cleary P.P. and Ferrieri P. 1991.C5a peptidase gene from group B streptococci. Genetics and Molecular Biology of Streptococci, Lactococci, and Enterococci 230-232,

13. Suvorov A.N., Tang C.E. and Ferretti J.J. 1992.Analysis of the streptolysin O (slo) gene of Streptococcus pyogenes. Proceedings of the IXth Landsfield International 10-14,

POTENTIAL PATHOGENIC PROPERTIES OF MEMBERS OF THE STREPTOCOCCUS MULLERIS GROUP

M.D.R. Willcox[1,2], M. Patrikakis[1], and K.W. Knox[1]

Summary

The *Streptococcus milleri* (SMG) group have been shown to posssess *in vitro* traits that may be involved in pathogenesis. Members of *Streptococcus constellatus* produced an enzyme with thrombin-like activity that was located on the cell wall. Lancefield group C SMG were able to aggregate rat platelets, with a bacterial cell-surface protein mediating the reaction. All SMG strains were able to bind fibronectin via a cell-surface protein. The *in vitro* traits were not correlated with each other, suggesting that SMG strains possess a wide variety of pathogenic traits which may be involved in the production of abscesses or endocarditis

---

1 Institute of Dental Research, 2 Chalmers Street, Surry Hills, NSW 2010, Australia
2 Cooperative Research Centre for Eye Research and Technology, PO Box 1, Kensington, NSW 2033 Australia (corresponding author)

INTRODUCTION

The *Streptococcus milleri* group (SMG) consists of the species *Streptococcus anginosus, Streptococcus intermedius* and *Streptococcus constellatus* [1,2,3]. Members of the SMG have been shown to produce purulent abscesses in many areas of the body including the mouth, brain and spleen [4,5], account for 9-15% of endocarditis caused by streptococci [4] and have been shown to produce caries in gnotobiotic rats [6]. Although able to cause these severe disease states, SMG strains are considered to be normal flora of the human oral cavity, upper respiratory and gastrointestinal tracts [4,5,7]. It has been shown that the three species are found associated with different diseases within the body; for example *S.intermedius* is more likely to be isolated from dental plaque than other species and strains causing urogenital infections are usually members of *S.anginosus* [2].

SMG strains isolated from abscesses adhere in greater numbers to buccal epithelial cells, and strains isolated from infections in the body (including abscesses) are able to bind more fibronectin than other strains [8]. The amount of fibronectin bound is related to the ability to adhere to saliva-coated hydroxyapatite [8]. In addition, strains can produce enzymes that may hydrolyse connective tissue components [2]. Recent data have demonstrated that strains possessing Lancefield group C polysaccharides are able to bind large amounts of albumin [9] and this may be related to their ability to aggregate human platelets.

The aim of the present study was to determine the location of degradative enzymes in the cells of SMG strains and to examine strains for their ability to bind to rat fibronectin and aggregate rat platelets.

MATERIALS AND METHODS.

Bacterial strains.

Strains chosen belonged to each of the three species in the SMG and certain strains within each species cross-reacted with antisera to Lancefield polysaccharides. *S.constellatus* strains (with Lancefield group) were AM 699 (-), NCTC 10714 (-), NCTC 11063 (-) and POW 5 (F), *S.intermedius* strains were UNS 35 (-), POW 3 (-) and C5 (C) and *S.anginosus* strains were NHM 10 (A), G5:3 (-), POW 1 (C), PC 4890 (C) C117221-1 (C) and F4 (F).

Localisation of Hydrolytic Enzymes.

The arylamidase substrates used were: *N-tert*-butoxycarbonyl (BOC)-Val-Pro-Arg-7-Amido-4-methylcoumarin (AMC; a synthetic substrate for thrombin-like activity) and BOC-Lys-AMC (has previously been shown to be hydrolysed by certain viridans streptococci [10]). Substrates were dissolved in a minimum of dimethylsulphoxide and diluted in 50mM *N*-tris (hydroxymethyl) methyl-2-aminoethenesulphonic acid buffer (pH 7.5; TES) to 100μg/ml. A suspension of whole bacterial cells (suspended in TES to an optical density at 620nm of 0.1; 50μl) was mixed with 20μl of substrate and incubated for 24h at 37°C. Substrate degradation was visualised by measuring the release of the fluorogenic compound under ultra violet light.

For the fractionation of cells, cultures were grown in Brain-Heart Infusion (BHI) for 18h at 37°C. Cell walls were obtained by digestion of whole cells (in the pellet from the above centrifugation step) with 20μg/ml mutanolysin (1h, 37°C) in 20mM Tris/HCl buffer (pH 6.8) containing 40% raffinose. Raffinose was added to maintain protoplast integrity during mutanolysin digestion [11]. The protoplasts were removed by centrifugation (9,000g, 10min) and the supernatant, containing the solubilised cell wall proteins, stored at 4°C until required. Cell membranes were then removed from the protoplasts. After centrifugation, the protoplasts were lysed by the addition of 40mM Tris/HCl (pH 6.8), shaken for 10 seconds and the cytoplasmic fraction (supernatant) and cell membrane fraction (pellet) separated by centrifugation (12,000g, 30min). The cytoplasmic fraction was removed. The cell membrane fraction was washed 4 times in ice-cold 20mM potassium acetate buffer (pH 7.0) containing 2.5mM $MgCl_2$ and finally resuspended in 1ml 50mM Tris/HCl (pH 8.6) containing 4% (v/v) Triton X100 [12]. After incubation at ambient temperature for 2h, the mixture was centrifuged (12,000g, 30min) and the supernatant containing solubilised cell membrane proteins removed and stored at 4°C.

After fractionation of cells, the total protein was estimated using a commercially available Lowry assay (Sigma). A standard curve of free 4-methylcoumarin was constructed (from 1mmoles to 0.01mmoles) and the absorbance read at 343nm. Substrates tested were the 4-methylcoumarin derivatives of lysine and valine-proline-glutamic acid; hydrolysis of the latter indicates a thrombin-like activity. Substrates (100μg/ml) and 0.1mg protein (cell wall or membrane fractions) were mixed incubated for 24h. The amount of released 4-methylcoumarin measured and the reactivity of the cell fractions was expressed as the amount of 4-methylcoumarin released per mg protein.

Binding to Rat Fibronectin

The assay used to measure the ability of strains to bind 1μg fibronectin (Fn) has been described previously [8]. ^{125}I-labelled Fn was allowed to bind to washed cell suspensions (5x10^{8} cells, resuspended in PBS containing 1% bovine serum albumin and 0.05% Tween 20) for 30min at 37°C. The cells were then centrifuged and the amount of radioactivity associated with the cell pellet measured. Cells were also tested for fibronectin binding after digestion of the cell surface with pronase (1mg/ml; Sigma) for 1h at 37°C, with subsequent extensive washing in buffer to remove the enzyme.

Hydrophobicity

A standard method [13] was used to measure the relative hydrophobicities of the strains. Washed cells (3.0ml) were mixed for 60s with 200μl hexadecane. The percentage hydrophobicity was measured as the decrease in optical density of the aqueous phase compared to controls with no hexadecane added.

Aggregation of Rat Platelets

Six were selected to be tested for their aggregation of platelets on the basis of differences in thrombin-like activity and fibronectin binding. Washed bacteria were concentrated fourfold giving a standard suspension of 4 x 10^9 cells/ml. Blood was obtained from rats by exsanguination of the heart under anaesthetic. Fresh blood was mixed immediately with 0.1M trisodium citrate (9:1[vol/vol]) and centrifuged at 100g for 20min at 20°C. The supernatant (platelet rich plasma; PRP) was removed and the remaining blood centrifuged at 3,000g for 10min to give platelet poor plasma (PPP). PRP was adjusted using PPP to give an $O.D._{660}$ of 0.5. A standard reaction mixture of 0.25ml fresh PRP and 0.25ml bacterial suspension were tested at 37°C with controlled stirring on a recording aggregometer (Chrono-log Corp.). After a baseline with PRP was established, the bacterial suspension was added to the cuvette. ADP was added to PRP suspensions as a positive control of normal platelet aggregation and also after 22 minutes of bacteria/platelet incubation to test the degree of aggregation seen. Strains were also tested for platelet aggregating ability after protease treatment of the cell surface (see above).

RESULTS

Localisation of Arylamidase Activity.

All strains of each species produced lysine-arylamidase activity, whereas only strains of *S.constellatus* produced thrombin-like activity. Lysine-arylamidase activity was localised to the cell membrane fraction for all species and showed a narrow range of activity, 1291µmoles/mg protein ± 196. Thrombin-like activity for *S.constellatus* strains was localised to the cell wall protein fraction and showed a range of activity from 1355μmoles/mg protein to 2432μmoles/mg protein.

Binding of Fibronectin to Strains and the Effect on Cell-Surface Hydrophobicity

Table 1 demonstrates the ability of strains to bind to rat fibronectin. Strains within each species showed a wide variation with *S.anginosus* F4 showing the greatest level of binding, 198 moleucles/cell, and *S.anginosus* NHM 10 binding only 12 molecules/cell. Pronase treatment of the cell surface resulted in significant reductions in fibronectin binding (table 1), most strains showed >80% reduction in fibronectin binding and all showed ≥45% reduction (table 1).

The cell surface hydrophobicity of all strains was increased by 3-36% after binding fibronectin. However, the amount of fibronectin binding did not correlate with the increase in hydrophobicity. For example, *S.anginosus* F4 bound the most fibronectin and gave a 36% increase in cell-surface hydrophobicity when fibronectin was bound to the cell surface, whereas *S.intermedius* C5 gave the second highest binding to fibronectin but the cell-surface hydrophobicity only increased 5% in the presence of fibronectin.

Aggregation of Rat Platelets

All strains showed some reaction when added to platelet rich plasma (figure 1). After the addition of ADP at 22 minutes, an increase in platelet aggregation was seen with strains, POW 5, POW 3 and F4, that gave the lowest values.

Protease treatment eliminated the platelet/bacterial interaction for all strains. A control of ADP, added after 22 minutes, was able to aggregate platelets, indicating the platelets were still in a functional state.

DISCUSSION

This study has shown that the SMG display a range of characteristics that may be important pathogenic determinants in the formation of both abscesses and endocarditis. Previous studies have demonstrated associations between certain types of SMG and diseases. For example, *S.intermedius* strains are commonly isolated from brain abscesses and hyaluronidase producing strains are more commonly isolated from abscesses [14]. The production of thrombin-like activity may indicate the possession of another biochemical characteristic that is correlated with disease. Thrombin converts fibrinogen to fibrin and is one of the major enzymes in the blood coagulation pathway. Coagulation of plasma may function in the production of abscesses and endocarditis as a primary step in the sealing off of bacteria in a platelet/fibrin clot. This process is thought to be beneficial to the host, as the pathogens are contained in defined areas. However, recent evidence suggests that the production of bacterial/platelet/fibrin clots may also be beneficial to the bacteria. Phagocytes isolated from abscesses have been shown to be ineffective in killing bacteria [15]. Also, the production of platelet/fibrin clots surrounding bacteria may effectively isolate the bacteria from other aspects of the host defence system.

The ability to bind fibronectin has previously been shown to be related to adhesion of SMG to saliva-coated hydroxyapatite and the isolation of strains from infections [8]. It is now clear that the binding of fibronectin is mediated by proteins on the cell-surface and that fibronectin binding increases the cell-surface hydrophobicity of strains to varying extents. The ability of *S.aureus* strains to bind fibronectin is correlated with their ability to invade the body [16] and fibronectin associated with fibrin clots promotes the adhesion of *S.aureus* to the clot [17]. Mutants of *S.aureus* showing reduced binding to fibronectin also show reduced adhesion to rat heart valves [18]. For the viridans streptococci, it has been shown that inactivation of fibronectin binding by transposon mutagenesis results in a reduced virulence of strains in the rabbit model of endocarditis [19]. Furthermore, for group C streptococci isolated from wounds, fibronectin or the C-terminal domain, was critical for adhesion to fibrin clots [20]. These observations may be important in the ability of strains to cause infections or be phagocytosed. It has been demonstrated that bacteria coated with fibronectin are more easily phagocytosed by polymorphonuclear leucocytes or macrophages [21,22]. Thus the binding of fibronectin to bacteria may have dual roles, allowing the bacteria to adhere to surfaces and also promoting the ingestion of bacteria by host defence cells.

Platelet aggregation may be a prerequisite for the production and progression of abscesses or endocarditis. In a previous study, it was demonstrated that the SMG cross-reacting with Lancefield group C antisera were the only strains able to aggregate human platelets [9]. The current study demonstrated that all strains reacted to some extent with rat platelets. However, only strains of Lancefield group C showed complete aggregation as determined by the addition of ADP at 22 minutes. The other strains either incompletely aggregated platelets or caused platelet agglutination only, as ADP increased the aggregation response. Thus, the SMG types that aggregate rat platelets are the same types that aggregate human platelets. The platelet aggregation response appeared to be due to a cell-surface protein on the SMG, as protease treatment of cells prevented aggregation. Platelet aggregation did not correlate with the ability of cells to bind fibronectin or the production of thrombin-like activity.

CONCLUSION

The SMG possess a number of cell-surface characteristics that may be related to their ability to produce abscesses and endocarditis. Thrombin-like activity was demonstrated using a synthetic fluorescently labelled substrate and was expressed by *S.constellatus* strains only. Only strains cross-reacting with Lancefield group C antisera showed complete aggregation of rat platelets. All strains were able to bind rat fibronectin, the binding being mediated by a cell-surface protein.

ACKNOWLEDGMENTS

This work was supported by grants from the Australian National Health and Medical Research Council of Australia and the Australian Dental Research Fund Inc.

Table 1. Binding of fibronectin to SMG strains and the effect on cell-surface hydrophobicity

Species	Strain	Fibronectin binding (molecules per cell)	Effect of protease fibronectin binding (% decrease)	Change in cell -surface hydrophobicity (% increase)
S.constellatus	AM 699	25	80	9
	NCTC 10714	23	70	21
	NCTC 11063	62	81	9
	POW 5	31	45	14
S.intermedius	UNS 35	56	93	5
	POW 3	57	81	29
	C5	105	97	5
S.anginosus	NHM 10	12	92	3
	G5:3	24	92	4
	POW 1	96	94	7
	PC 4890	70	93	13
	C117 221-1	74	88	4
	F4	198	97	36

ll standard deviations were less than 10% of the mean.

Figure 1. Aggregation of Rat Platelets by SMG Strains

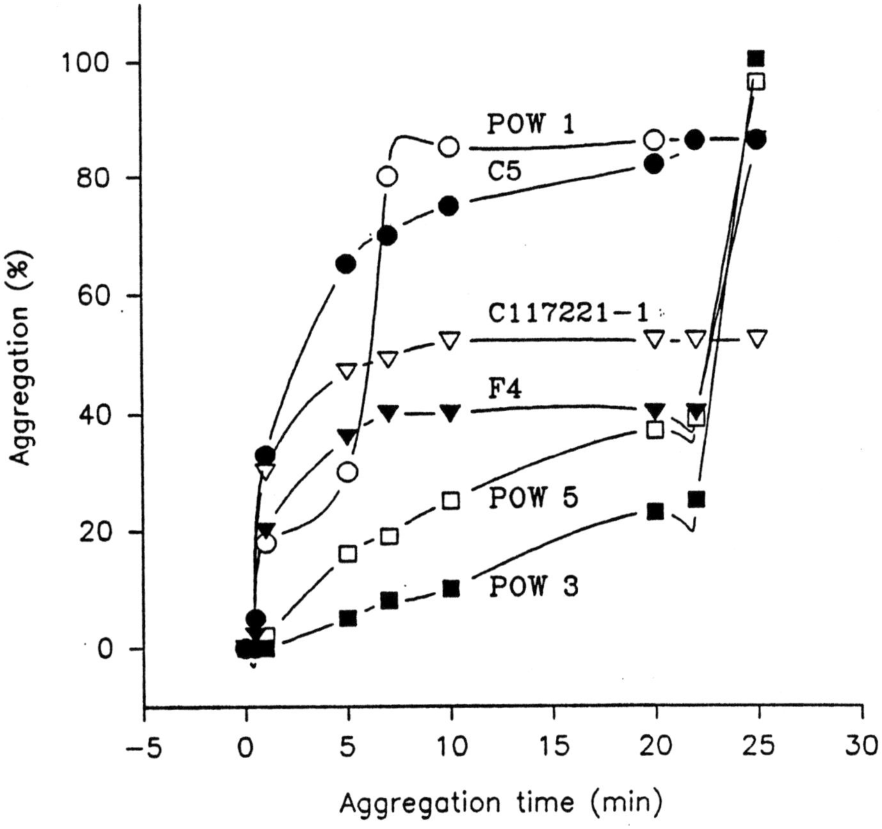

REFERENCES

1. R.A.Whiley and J.M.Hardie. "DNA-DNA hybridisation studies and phenotypic characteristics of strains within the '*Streptococcus milleri*' group". *Journal of General Microbiology* Vol.135. 1989. p.2623.
2. R.A.Whiley, H.Fraser, J.M.Hardie and D.Beighton. "Phenotypic differentiation of *Streptococcus intermedius, Streptococcus constellatus,* and *Streptococcus anginosus* strains within the '*Streptococcus milleri* group'". *Journal of Clinical Microbiology* Vol.28. 1990. p.1497.
3. R.A.Whiley and D.Beighton. "Emended descriptions and recognition of *Streptococcus constellatus, Streptococcus intermedius* and *Streptococcus anginosus* as distinct species". *International Journal of Systematic Bacteriology* Vol.41. 1991. p.1.
4. J.Gossling. "Occurrence and pathogenicity of the *Streptococcus milleri* group". *Reviews of Infectious Diseases.* Vol.10. 1988. p.257.
5. K.L.Ruoff. "*Streptococcus anginosus* ("*Streptococcus milleri*"): The unrecognised pathogen". *Clinical Microbiology Reviews* Vol.1. 1988. p.102.
6. D.B.Drucker and R.M.Green. "Relative cariogenicities of *Streptococcus milleri* and other viridans streptococci in gnotobiotic hooded rats". *Archives of Oral Biology.* Vol.26. 1978. p.625.
7. B.Mejare and S.Edwardsson. "*Streptococcus milleri* (Guthof); an indigenous organism of the human oral cavity". *Archives of Oral Biology.* Vol.20 1975. p.757.
8. M.D.P.Willcox and K.W.Knox. "Surface-associated properties of

Streptococcus milleri group strains and their potential relation to pathogenesis". *Journal of Medical Microbiology* Vol.31. 1990. p.259.

9. M.D.P.Willcox, M.Patrikakis, C.Y.Loo and K.W.Knox. "Albumin-binding protein on the surface of the *Streptococcus milleri* group and characterisation of the albumin receptor of *Streptococcus intermedius* C5. *Journal of General Microbiology*. In Press. 1993.

10. K.A.Homer, R.A.Whiley and D.Beighton. "Proteolytic activity of oral streptococci". *FEMS Microbiology Letters*. Vol.67. 1990. p.257.

11. J.L.Siegel, S.F.Hurst, E.S.Liberman, S.E.Coleman and A.S.Bleiweis. "Mutanolysin-induced spheroplasts of *Streptococcus mutans* are true protoplasts". *Infection and Immunity*. Vol.31. 1981. p.808.

12. E.Floderus and L.E.Linder. "Localisation of aminopeptidases in *Streptococcus sanguis* strain 903". *Current Microbiology*. Vol.20. 1990. p.255.

13. M.Rosenberg, D.Gutnick and E.Rosenberg. "Adherence of bacteria to hydrocarbons: a simple method for measuring cell surface hydrophobicity". *FEMS Microbiology Letters*. Vol.9. 1980. p.29.

14. P.F.Unsworth. "Hyaluronidase production in *Streptococcus milleri* in relation to infection". *Journal of Clinical Pathology*. Vol.42. 1989. p.506.

15. J.J.Finlay-Jones, P.H.Hart, L.K.Spencer, M.F.Nulsen, P.A.Kenny and P.J.McDonald. "Bacterial killing *in vitro* by abscess-derived neutrophils". *Journal of Medical Microbiology*. Vol.34 1991. p73.

16. R.A.Proctor, G.Christman and D.F.Mosher. "Fibronectin-induced agglutination of *Staphylococcus aureus* correlates with invasiveness".

Journal of Laboratory and Clinical Medicine. Vol.104. 1984. p.455.

17. P.T.C.Y.Toy, L.-W.Lai, T.A.Drake and M.A.Sande. "Effect of fibronectin on adherence of *Staphylococcus aureus* to fibrin thrombi in vitro". *Infection and Immunity*. Vol.48. 1985. p.83.

18. J.M.Kuypers and R.A.Proctor. "Reduced adherence to traumatised rat heart valves by a low-fibronectin-binding mutant of *Staphylococcus aureus"*. *Infection and Immunity*. Vol.57. 1989. p.2306.

19. J.H.Lowrance, L.M.Baddour and W.A.Simpson. "The role of fibronectin binding in the rat model of experimental endocarditis caused by *Streptococcus sanguis"*. *Journal of Clinical Investigation* Vol.86. 1990. p.7.

20. G.S.Chhatwal, P.Valentin-Weigand and K.N.Timmis. "Bacterial infection of wounds: fibronectin-mediated adherence of group A and C streptococci to fibrin thrombi in vitro". *Infection and Immunity* Vol.58. 1990. p.3015.

21. K.D.Yang, N.H.Augustine, L.A.Gonzalez, J.F.Bohnsack and H.R.Hill. "Effects of fibronectin on the interaction of polymorphonuclear leucocytes with unopsonized and antibody-opsonized bacteria". *The Journal of Infectious Diseases* Vol.158. 1988. p.823.

22. J.L.Kluftinger, N.M.Kelly, B.H.Jost and R.E.W.Hancock. "Fibronectin as an enhancer of nonopsonic phagocytosis of *Pseudomonas aeroginosa* by macrophages". *Infection and Immunity* Vol.57. 1989. p.2782.

PLASMIDS OF BACILLI SOIL STRAINS

V.Z. Nezameydinova, A.Sh. Kanapina, A.A. Prozorov

ABSTRACT

Strains from bacilli collection, related to Bacillus subtilis 168 were examined for the presence of plasmid DNA. 21 out of 32 tested strains were shown to harbor plasmid DNA. These strains carry 1 or more plasmids of 5-10 kb; two of them carried large plasmids. Isolated plasmids probably do not bear antibiotic resistance markers. All of them except two large ones displayed various degrees of similarity to each other and were homologous to pL512 and pBN27. Homology to other known plasmids of B. subtilis or Staphylococcus was not observed.

The restriction map was constructed for the 8 kb cryptic plasmid p1414, and it was labeled by selectable Cm^R marker. As is known the plasmid replicates by the "rolling circle" mechanism. It was shown that p1414 plasmid and its derivatives are stable in the cells. The data obtained indicate the integration possibility of p1414 or its fragments into the host cell chromosome.

1. INTRODUCTION

Bacillus subtilis plasmid DNA was isolated and described long ago (1-3), but poorly used as a cloning system (4). Most of the previous vectors used for cloning in B. subtilis were the plasmids derived from S. aureus (5). These vectors are characterized by low stability caused by replication mechanism in bacilli cells (6,7). A search for natural Bacilli plasmid preferable for vector construction is still in progress. It seems most perspective to use the naturally isolated B. subtilis plasmids from the collection of Bacilli soil strains.

In the present work the main properties of cryptic plasmids isolated from the variety of soil Bacilli strains closely related to Bacillus subtilis 168 are described and the relationship between one another as well as between these plasmids and those of other bacteria assessed. The results of studying replication mechanism, stability and interaction with the host cell chromosome are also presented.

2. Materials and Methods

The strains and plasmids used in this work are listed in Table 1. Soil strains were isolated and characterized by Y.E. Kozlovsky (8).

The pGG3 plasmid was used as a cat gene source. In pGG3 cat-gene from pBD64 (9) was inserted in pUC19 polylinkers surrounded by intact restriction SalI sites.

Bacteria were grown on Nutrient Broth liquid media. The tests on antibiotic resistance were performed on NB agar, using the following antibiotics: chloramphenicol (5 ug/ml), kanamicin (30 ug/ml), eruthromicin (5 ug/ml), tetracycline (10 ug;ml), ampicillin (50 ug/ml) and riphampicin (1 ug/ml).

Transformation of B. subtilis component cells was performed by Anagnostophoulos and Spicizen method with modifications (10).

Plasmid DNA was isolated by Birnboim and Dolly method (11) with subsequent phenol purification. Single stranded DNA were obtained from crude lysates by Riele method (12). Riphampicin treatment of cells at 100 ug/ml concentration for ss-DNA accumulation during plasmid replication was carried out, using the method of Boe et al. (13). Plasmid copy number determination was performed according to the procedure described by Proyan (14). For preparative purposes plasmid DNA was purified by means of centrifugation to equilibrium in a CsCl gradient (11). Chromosomal DNA was isolated by the mini-extraction method (15).

Restriction, ligation and DNA treatment with Klenov fragment, as well as Bal31 and s 1 nucleases were carried out according to procedures described by Maniatis et al. (11). Restriction enzymes for molecular cloning were obtained from Boehringer Mannheim Biochemicals and used as indicated by manufacturer. Gel electrophoresis of plasmid DNA and DNA-fragments was performed in 0.8% agarose "Sigma". DNA fragments were isolated from low melting agarose "Sigma" (11). DNAs were transferred to nylon filters (Hybond-N, Amersham) and hybridized by using standard procedures (11).

Hybridization and washing of filters were carried out at 65 C as described by Maniatis et al. (11).

For determination of the plasmid stability all examined strains with plasmids were grown on NB medium under normal conditions during 100 generations. Samples taken at equal intervals were plated on NB agar and then on selective agar with chloramphenicol (5 ug/ml) to determine the percentage of Cm^R colony forming cells.

At every indicated time more than 500 colonies were tested on resistance to chloramphenicol, while 20 ones were tested on the presence and size of plasmid by electrophoresis.

Plasmid copy number was determined by densitometry of total DNA electrophoregrammes as described by Proyan (14).

3. Results

3.1 General Properties of Plasmids

In the present work the soil strain collection closely related to Bacillus subtilis 168 was used. These strains were isolated and characterized by Y.E. Kozlovsky (8). 32 strains were examined for the presence of plasmid DNA, out of which 21 were shown to contain one or more plasmids of different size. Two strains harbor heavy plasmids of more than 30 kb together with long ones. The presence of plasmids in these strains could not be associated with resistance to antibiotics in that they were sensitive to different antibiotics used in the test.

To examine the relationship among these plasmids, total DNA isolated as described by te Riele (12) was transferred by Southern blotting to Hybond-N membranes and hybridized with radioactively labelled plasmid DNA p1414 from Bacilli strain 1414. Strain 1431 which did not harbor plasmid was used as reference strain. Strong hybridization was seen with p1414 and all plasmids (Fig. 1,A). Plasmid DNA of p1414 was used as a probe in hybridization with a number of well characterized plasmids from Bacilli and Staphylococcus (Fig. 1): pUB110, pE194, pBC16, pSA0501, pLS12, pC194, pLS14, pT127, pLS1, pBN27. Hybridization signals were detected only from cryptic plasmid DNAs pLS12 (1) and pBN27 (16) (Fig. 1,B).

Plasmid p1414 was chosen for more detailed characterization. Plasmid is 8 kilibases in size and presents in cells at a copy number of 8-10 per chromosome equivalent. The restriction map of p1414 is shown in Fig. 2.

3.2 Replication Mechanism of p1414 Plasmid

The presence or absence of the intermediate single-stranded DNA plasmid forms is the main feature of plasmid replication type. It was observed that ss-DNA does not accumulate during p1414 replication under standard conditions (Fig. 3, lines 7,8). However, the appearance and accumulation of ss-DNA in cell lysate of 1414 strain (Fig. 3, lines 3,4) during cell treatment by high riphampicin concentration has been detected. Consequently, plasmid p1414 replicates by the "rolling circle" mechanism. Ss-DNA does not accumulate under ordinary conditions, but the riphampicin treatment repressing an action of RNA-polymerase results in ss-DNA accumulation. This indicates that RNA-polymerase is involved in conversion of ss-DNA into ds-DNA.

3.3. Mapping of p1414 Regions Responsible for Plasmid Replication

DNA p1414 was treated with different restriction enzyme combinations to realize the minimal replicon borders and resulting 2-3 large fragments were ligated with a fragment from pGG3 containing cat gene (1,1 kb). B. subtilis cells RUB 834 were transformed by the ligated mixture with further selection by Cm-resistance marker. The 5,9 kb PstI-KpnI fragment was isolated, providing the plasmid replication in B. subtilis cells. The resulting plasmid was named pVA2 (Fig. 2). Then PstI-KpnI fragment was shorted by Bal31 nuclease treatment. Resulting fragments were completed by Klenov fragment and ligated with fragment carrying cat gene from pGG3 (1,1 kb). Thus, the 3,7 kb plasmid, (Fig. 2) named pVA26 was produced. The plasmid successfully replicated in B. subtilis cells. The structures needed for replication (replication initiation point and gene encording replication protein) lie on the borders of 0,57 kb Bc1I-HindIII and 1,53 kb HindIII-HindIII fragments (Fig. 2).

The structures responsible for ss-DNA conversion into ds-DNA (ori-) lay on 2,1 kb fragment deleted in pVA22 plasmid (Fig. 2). During the replication of this plasmid ss-DNA accumulates without additional treatment (Fig. 3, lines 1,2 and 5,6). In this plasmid the conversion of ss-DNA into ds-DNA appears to break.

3.4 Determination of p1414 Plasmid Stability and its Derivatives

Plasmid p1414 was marked by cat gene to determine its stability. The fragment containing cat gene was inserted in p1414 SalI site and the resulting plasmid was named pVA14 (Fig. 2). The strains carrying pVA14 and pVA26 plasmids, as well as pBD64 as a control plasmid, were grown during 100 generations under nonselective conditions. PVA14 and pVA26 plasmids were kept in 100% of cell population. During this period pBD64 plasmid was lost completely. In addition pVA14 and pVA26 DNAs, isolated from clones after 100 generations under nonselective conditions didn't reveal any notable differences (by electrophoregramme, data not shown) in comparison with the native ones. Therefore, p1414 plasmid and its derivatives have high structural and segregational stability.

3.5 Interaction of p1414 Plasmid with Host Cell Chromosome

Plasmid p1414 and its derivatives were observed to hybridize with RUB 834 host strain chromosome. The bands of hybridization, different in mobility from linear p1414 plasmid were detected on radioautographe of DNA chromosome blot hybridization restricted by SalI restriction endonuclease with p1414 probe (Fig. 4). The DNA chromosome of native non-plasmid strain RUB 834 was not hybridized with p1414 probe (Fig. 4).

4. Discussion

From 10 to 60% of soil bacilli strains harbor plasmids (1-3, 17, 18), most of them being small (5-15 kb). Large ones (more than 20 kb) occur rarely (3, 17-19). About 60% of the tested strains contain small plasmids, and only 2 out of 32 ones contain heavy plasmids. All plasmids except two large ones were shown to be homologous or partly homologous with each other. Preliminary restriction analysis reveals that several parts of the plasmids are identical, and the rest of them have identical fragments. In addition, the plasmids in this work are homologous with pLS12 (1) and pBN27 (16). The data obtained in the present work agree with the reports of several authors (17-19) on wide distribution of related plasmids in bacilli strains. Effective mechanisms providing the keeping of natural bacilli plasmids in cell population are of interest for us. The data on wide distribution of cryptic plasmids in Bacilli suggest not only the possibility of horizontal plasmid transfer between the strains under natural conditions but also an adaptive value of plasmids for a bacterial cell.

One of the isolated plasmids p1414, 8 kb in size, was studied ore carefully, having 8-10 copies per chromosome equivalent. It was shown that the plasmid is homologous not with the known staphylococcus plasmids but with cryptic pLS12 (1) and pBN27 (16) ones. P1414 replicates by the "rolling circle" mechanism and host cell RNA-polymerase participates in the very effective conversion of ss plasmid DNA into ds-DNA. These results agree with the data on replication of pTA1060 (20,21) and pBAA1 (22). Plasmid pUB110 replicates by similar mechanism as well. Some authors relate this plasmid to bacilli origin, even though it was isolated from staphylococcus (23,24). Plasmid p1414 and its derivatives have high segregational and structural stability. These data also agree with the ones on cryptic plasmid properties (20-22).

The fact of p1414 interaction with host cell chromosome is very interesting. The multiple bands of chromosomal DNA hybridization with labelled plasmid can not be explained by structural changes in autonomously replicating plasmids, since the preparations of p1414 DNA plasmid and its derivatives were highly homogenous in restriction analysis.

The information on hybridization showed the multiple integration of p1414 or its fragments into the host cell chromosome. The similar data in Bacilli plasmid investigation were obtained for the first time. The studying of this phenomenon may reveal completely new properties in Bacilli plasmid biology.

5. CONCLUSION

More than 20 cryptic plasmids were isolated from Bacilli soil strains. One of the plasmids was mapped and its replication mechanism studied. High structural and segregational stability and integration possibility of this plasmid or its parts into the host cell chromosome was shown.

Our thanks are due to Y.E. Kozlovsky for permitting us to use his soil bacilli collection for plasmid search. We are also grateful to N.V. Alexeenko, G.G. Maslova, and K.N. Kroon for their help.

TABLE 1.

Strains and plasmids used in work.

Strain	Genotype	Source
RUB 834	trpC2 metB2 aroB2 hisH2 tyrA1	B.G.S.C.
32 strains of soil bacilli	prototrophes	Inst. of Gen. Genetics, (8) Kozlovsky Y.E.

Plasmid	Markers	Reference
pBD64	Km^RCm^R	Gryczan et al., 1980
pC194	Cm^R	Ehrlich, 1977
pUB110	Km^R	Gryczan et al., 1978
pE194	Em^R	Hourinouchi,Weisblum, 1982
pBC16	Tc^R	Bernhard et al., 1978
pLS12	-	Tanaka et al., 1977
pLS14	-	--//--
pBC1	Tc^R	
pLS1	Tc^R	Laks et al., 1986
pSA0501	Sm^R	Gryczan et al., 1986
pT127	Tc^R	Ehrlich et al., 1977
pBN27	-	Kim, 1989 (16)

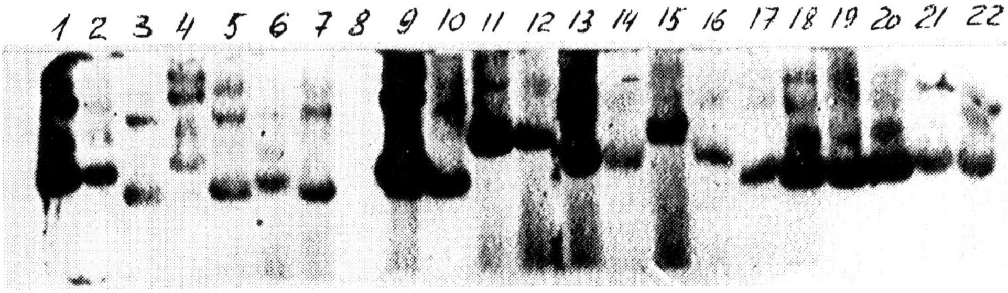

Figure 1a Results in p 1414 probe hybridization with studied bacilli strain plasmid DNAs. Number of strains: 1 track - strain 1801; 2 - 115; 3 - 865; 4 - 1315; 5 - 1385; 6 - 1387; 7 - 1410; 8 - 1431; 9 - 1434; 10 - 1437; 11 - 1516; 12 - 1525; 13 - 1414; 14 - 1546; 15 - 1553; 16 - 1570; 17 - 1668; 18 - 1674; 19 - 1698; 20 - 1776; 21 - 1870; 22 - 1878

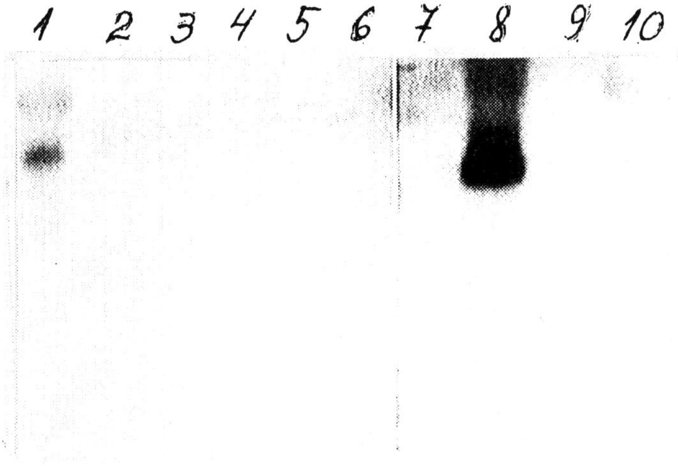

Figure 1b Results of p1414 probe hybridization with known bacilli and staphylococci plasmid DNAs: 1 - pBN27; 2 - pUB110; 3 - pLS1; 4 - pT127; 5 - pLS14; 6 - pC194; 7 - pSA0510; 8 - pLS12; 9 - pBC16; 10 - pE194.

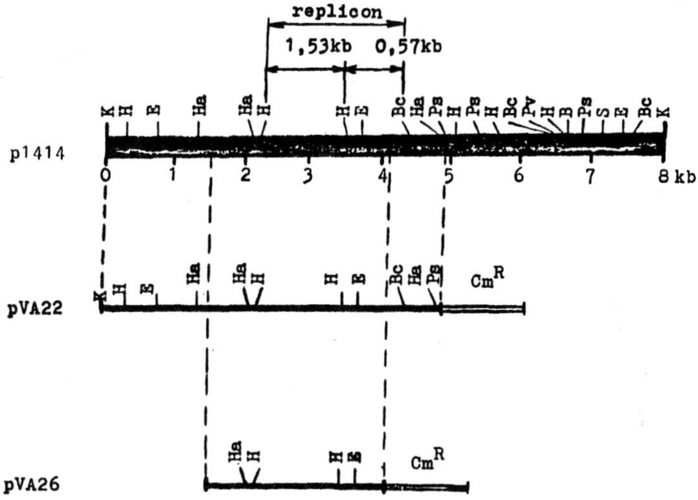

Figure 2 The restriction map of p1414 plasmid and its derivatives. The following restriction sites are indicated: E-EcoRI; B-BamHI; Bc-Bc1I; Ha-HaeIII; K-KpnI; Pv-PvuII; Ps-PstI; S-Sa1I. Cm^R - CAT-gene.

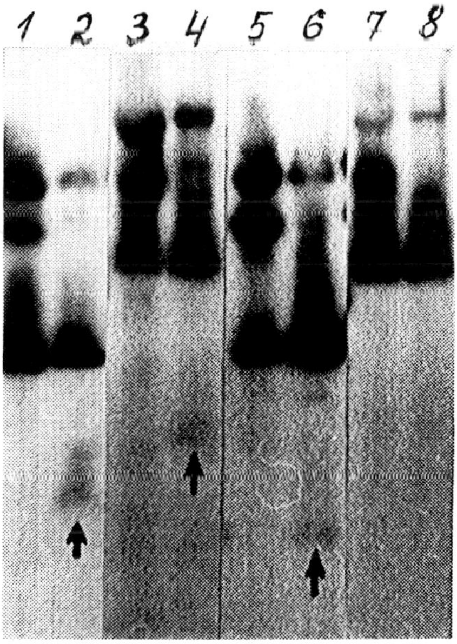

Figure 3 ss-DNA accumulation during p1414 and pVA22 plasmids replication. Samples treated by riphampicin 1 - pVA22 + S1 nuclease; 2 - pVA22; 3 - p1414 + S1 nuclease; 4 - p1414. Samples non-treated by riphampicin: 5 - pVA22 + S1 nuclease; 6 - pVA22; 7 - p1414 + S1 nuclease; 8 - p1414. S1 nuclease specifically degrades only ss-DNA. The positions of ss-DNA are indicated by arrows.

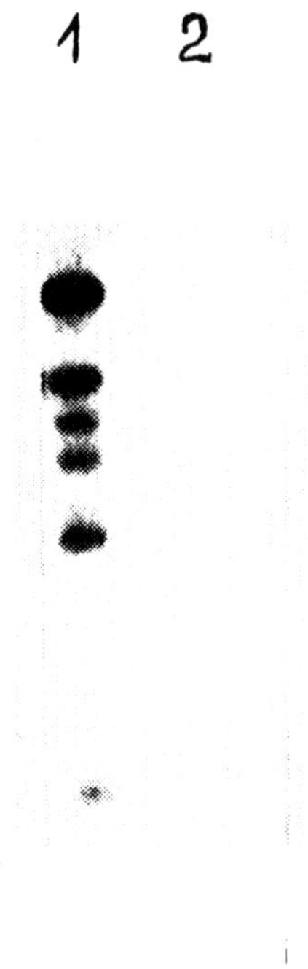

Figure 4 Results of p1414 probe hybridization with RUB 834 strain containing p1414 chromosomal DNA. Samples: 1 - chromosomal DNA of RUB 834 strain, containing p1414; 2 - chromosomal DNA of RUB 834 strain.

References

1. T. Tanaka, M. Kuroda and K. Sakaguchi. "Isolation and Characterization of Four Plasmids from Bacillus subtilis". J. of Bacteriol., Vol. 129, No. 3, 1977, pp. 1587-1494.

2. P.S. Lovett, E.I. Duvall and K.M. Keggins. "Bacillus Pumilus Plasmid pPL10: Properties and Insertion into Bacillus subtilis 168 by Transformation". J. of Bacteriol., Vol. 127, No. 2, 1976, pp. 817-828.

3. P.S. Lovett, M.G. Bramucci. "Plasmid Deoxyribonucleic acid in Bacillus subtilis and Bacillus Pumilus". J. of Bacteriol., Vol. 124, No. 1, 1975, pp. 484-490.

4. J. Kreft, C. Hugnes. "Cloning Vectors Derived from Plasmids and Phages of Bacillus". Current Topics on Microbiol. and Immunol., Vol. 96, 1982, pp. 1-17.

5. S.D. Ehrlich, B. Niaudet and B. Michel. "Use of Plasmids from Staphylococcus Aureus for Cloning of DNA in Bacillus subtilis". Current Topics on Microbiol. and Immunol., Vol. 96, 1982, pp. 18-29.

6. S.D. Ehrlich, H. te Riele, M.A. Petit, L. Janniere, Ph. Noirot and B. Michel. "DNA Recombination in Plasmids and the Chromosome of Bacillus subtilis". In: Bacillus Molecular Genetics and Biotechnology Application, 1986, pp. 27-33.

7. L. Janniere, C. Bruand and S.D. Ehrlich. "Structurally Stable Bacillus subtilis Cloning Vectors". Gene, Vol. 87, 1990, pp. 53-61.

8. Y.E. Kozlovsky, A.A. Prozorov. "The Restriction-Modification System in Bacilli Strains Related to Bacillus subtilis". "Dokl. of A.S.", Vol. 258, 1981, p. 1457-1459.

9. T. Gryczan, A.T. Shivakumar and D. Dubnau. "Characterization of Chimeric Plasmid Cloning Vehicles in Bacillus subtilis". J. of Bacteriol., Vol. 141, 1980, pp. 246-255.

10. C. Anagnostopoulos, J. Spizizen. "Requirements for Transformation in Bacillus subtilis". J. of Bacteriol., Vol. 81, 1961, No. 5, pp. 741-746.

11. T. Maniatis, E.E. Fritsch and J. Sambrook. Molecular Cloning. A laboratory manual, Cold Spring Harbor Laboratory, 1982, 479 pp.

12. H. te Riele, B. Michel and S.D. Ehrlich. "Single-stranded Plasmid DNA in Bacillus subtilis and Staphylococcus aureus". Proc. Natl. Acad. Sci. of USA, Vol. 83, 1986, pp. 2541-2545.

13. J. Boc, M. Gros, H. te Riele, S.D. Ehrlich and A. Gruss. "Replication Origins of Single-stranded DNA Plasmid pUB110". J. of Bacteriol., Vol. 171, 1989, pp.3366-3372.

14. S.J. Projan, S. Carllton and R.P. Novick. "Determination of Plasmid Copy Number by Fluorescence Densitometry". Plasmid, Vol. 9, No. 2, 1983, pp. 182-190.

15. L. Janniere and M.A. Petit. "Integration and Amplification of Foreign Genes in the Chromosome of Bacillus subtilis". In: EMBO Course on DNA Cloning in Bacillus subtilis. S.D. Ehrlich (Ed), Paris: Institute "Jacques Monod", 1985, pp. 39-40.

16. E.L. Kim. "pBN27 - the Cryptic Plasmid of Bacillus natto". Abstracts of conference "The Structure and Function of Biopolymers", Lvov, 1989, p. 30.

17. E. de Rossi, P. Brigidi, G. Riccardi, D. Matteurri. "Plasmid Screening in Termophilic Bacillus: Physical Characterization and Molecular Cloning", Carrent Microbiol., Vol. 19, 1989, pp. 13-19.

18. C. Parini, M.G. Fortina, P. LpManachini, E. de Rossi and G. Riccardi. "Detection and Characterization of Naturally Occurring Plasmids in Bacillus Licheniformis", FEMS Microbiol. Lett., Vol. 81, 1991, pp. 329-334.

19. K. Bernhard, H. Schrempf and W. Goebel. "Bacteriocin and Antibiotic Resistance Plasmids in Bacillus Cereus and Bacillus subtilis". J. of Bacteriol., Vol. 133, 1978, pp. 897-903.

20. Y. Uozumi, A. Ozaki, T. Beppu and K. Arima. "New Cryptic Plasmid of Bacillus subtilis and Restriction Analysis of other Plasmids Found by General Screening". J. of Bacteriol., Vol. 142, 1980, pp. 315-318.

21. S. Bron, P. Bosma, N. van Belkum and E. Luxen. "Stability Function in the Bacillus subtilis Plasmid pTA1060". Plasmid, Vol. 18, 1987, pp. 8-16.

22. K.M. Devine, S.T. Hogan, D.M. Higgins and D.Y. McConnell. "Replication and Segregational Stability of Bacillus Plasmid pBBA1". J. of Bacteriol., Vol. 171, 1989, pp. 1166-1172.

23. I.E. Maciag, J.F. Viert and J.C. Alonso. "Replication and Incompatibility Properties of Plasmid pUB110 in Bacillus subtilis". Mol. Gen. Genet., Vol. 212, 1988, pp. 232-240.

24. H. Leonhardt. "Identification of a Low-copy-number Mutation within the pUB110 Replicon and its Effection Plasmid Stability in Bacillus subtilis". Gene, Vol. 94, 1990, pp. 121-124.

PART 2

MOLECULAR BIOLOGY OF THE GRAM-NEGATIVE BACTERIA

Biological Properties of Salmonella Enteritidis which Circulated in Russia in 1985-1992

L.A. Kaftyreva, A.V. Zabrovskaya, E.V. Koitenkova, M.J. Nesterova, G.K. Degteva, R.V. Ermolina

Summary

Since 1985 the growth of S. enteritidis among humans has been noted in Russia. The factors of transmission were poultry meat and egg products. S. enteritidis phage type 1 dominated during all years of observation (1985-1992) and accounted for 95.6%.

More than 90% of the strains were sensitive to antibiotics, 8.3% were resistant from 1 to 6 markers.

Almost all strains had serotypospecific plasmid 38MD. Multiresistant ones had conjugative R-plasmid with 96 or 140 MD molecular weight additionally.

Restriction analysis of serotypospecific plasmids in 4 strains showed their identity in quantity of patterns and their molecular weights. 10 strains produced group I colicin.

The spectrum of the outer membrane proteins among S. enteritidis phage type 1 strains was practically identical irrespective of the source and year of isolation, antibiotic sensitivity and colicin production.

In conclusion it may be said that wide circulation of S. enteritidis phage type 1 among humans formed the variety of biological features coded by plasmids (antibiotic resistance, colicin production).

1. Introduction

Since the late 80ies in the European part of Russian the growth of salmonellosis has been registered. In the following years the considerable growth of outbreaks and sporadic cases of salmonellosis began in its other regions. As a whole, in each region morbidity growth was followed by the change of leading serovars. Before 1985 the S. typhimurium held the leading position and in the following years it was held by S. enteritidis [1].

Before 1985 S. enteritidis was among 10 serovars circulating with the greatest frequency among humans and animals. But its specific gravity did not prevail 1% [2].

In recent years poultry products, massively contaminated by microorganisms acquired main factors of S. enteritidis transmission.

Our further microbiological investigations showed that more than 70% of S. enteritidis strains isolated from patients belonged to phage type 1. Strains of this phage type were more frequently isolated from poultry, meat and egg products [3].

2. Materials And Methods

2.1 Strains

The investigations were based on studying biological properties of 1167 S. enteritidis phage type 1 strains isolated from patients (children and adults) in 1985 - 1992 on 25 different territories of Russia.

2.2. Biochemistry and Serological Identification

Cultural-morphological properties, antigen structure and fermentation activity were defined by traditional methods [4].

2.3 Phage Typing

We used the phages and the scheme for S. enteritidis phage typing, elaborated in National Institute of Hygiene, Budapest, Hungary [5].

2.4 Resistance Typing

All strains were tested on their susceptibility to seven antibiotics: chloramphenicol, kanamycin, streptomycin, tetracycline, polymixine, ampicillin and gentamicin by agar diffusion test with antibiotic disks. Minimal inhibition concentration was determined by the method of consecutive delutions.

2.5 Plasmid Analysis

Plasmid profile was determined in 567 isolates of S. enteritidis. The strains were grown overnight in nutrient agar plates and plasmid DNA was extracted by the method of Kado and Lui [6]. Samples were electrophoresed in vertical 0.8% agarose gels, stained with ethidium bromide and photographed under UV light. The molecular weight of plasmids was estimated by the migration distances and compared with that of standard plasmids from E. coli I53 R40a (96MD) and E. coli I53 RP4 (36MD).

Restriction analysis was conducted with endonuclease Hind 3 made by Dr. Arai (Department of Microbiology, Meiji College of Pharmacy, Setagaya-ku, Tokyo 154, Japan).

2.6 Colicin Production

Colicin production was defined in all strains of S. enteritidis phage type 1.

Test strains were cultured in nutrient plates and killed by chlorophorm vapor. 3 ml of soft agar containing expotentially growing indicator strains were poured in the plates and cultured at 37 C overnight. Colicin production was judged by the presence of the inhibition zone growth around the test cell culture [7].

2.7 Outer Membrane Protein Profiles

Outer membrane proteins were extracted by the modified method [8].

Outer membrane protein profiles in 17 strains of S. enteritidis phage type 1 were analyzed in polyacrilamide gels and compared. The strains differed by years of isolation, susceptibility to antibiotics (sensitive and resistant to 1, 2 and 5 antibiotics), and single strained produced colicin belonged to group 1.

2.8 Enterotoxin Production

Entertoxins (thermolabile LT and thermostabile ST) were studied in 15 strains on experimental models by the intragastric test on infant mice [9] and by the test of white mouse foot paw oedema [10].

3. Results

3.1 Biochemical and Serological Properties

Fermentation and serology of all studied S. enteritidis strains were typical to salmonella genus and didn't give difficulties during isolation from patients. They had stable fermentative features.

3.2 Phage Typing

88.4% of studies S. enteritidis strains belonged to phage type 1. Yearly distribution of S. Enteritidis phage types is presented in Table 1.

S. enteritidis phage type 1 dominated during all years of observation. Specific gravity of this phage type ranged from 82.2% (1987) to 95.6% (1985).

Strains of S. enteritidis phage type 1 were isolated from patients on 25 territories and dominated in all of them. They formed from 66% to 100% of studied strains. 7.5% of the strains belonged to other 10 phage types (2, 3, 4, 6, 7, 9, 10, 12, 16, 17). Each of these phage types was presented by 5 - 27 cultures. Phage types 9 and 16 were presented by 1 strain. 1.7% of the cultures didn't lay into the scheme of phage typing. 2.4% of the strains didn't lysate by any of 8 typical phages.

In Russian the circulation of 9 different phage types was determined. Phage type 1 predominated on all territories. The frequency of circulation in other ones was limited. Phage types 2, 10, 12 were isolated on two territories, the strains oh phage types 3 and 6 circulated on 5 territories, and those of phage type 4 - on six ones. Untypable strains were isolated on all territories.

3.3. Antibiotic Sensitivity

More than 90% of S. enteritidis strains oh phage type 1 were sensitive to antibiotics. MIC of gentamicin and polymixine didn't prevail 1 mkg/ml.

Chloramphenicol, kanamycin, tetracycline, streptomycin and ampicillin formed 2 - 8 mkg/ml. 97 strains (8.3%) were resistant to antibiotics (from 1 to 6 markers). MIC of antibiotics among such strains were: tetracycline - 300 mkg/ml, streptomycin - 400 mkg/ml, other antibiotics - 500 mkg/ml and more.

The strains, resistant to antibiotics, began to circulate from 1986. Markers of antibiotics resistance are shown in Table 2.

We studied 7 antibiotics (cloramphenicol, kanamycin, streptomycin, tetracycline, polymixine, ampicillin, gentamicin). Resistance was to 6. There was no resistance to polymixine. With more frequency circulated strains with 1 marker (Cm, Str, Km, Ap, Tc) - 4.5% to 2 antibiotics were 1.4% resistant. The clones multiresistant to antibiotics (4 and 6 markers of resistance) were isolated in 2.4%.

If the strains resistant to 1 - 2 antibiotics were isolated during all years of observation, beginning from 1986, multiresistant ones were isolated in 3 focuses in 1986 and 1988 during the intrahospital infection, and in each outbreak there was its own spectrum of resistance: Cm, Str, Km, Ap: Cm, Str, Km, Tc and Cm, Km, Str, Tc, Ap, Gm.

3.4 Plasmid Analysis

The analysis of 564 cultures with plasmid profile, out of which 518 were sensitive and 46 resistant to antibiotics, showed that all strains excluding 2 sensitive ones to antibiotics had plasmid with 38MD molecular weight, and multiresistant strains had additional conjugative R-plasmids with 96MD (markers of resistance Cm, Str, Km, Ap) and 140 MD (markers of resistance Cm, Str, Tc, Ap, Gm) molecular weight.

We could not discover R-plasmids among the strains resistant to one or two antibiotics.

The restriction analysis of serotypospecific plasmids (38MD), carried out by digestion with endonuclease Hind 3 of 4 phage type 1 S. enteritidis, circulating in St. Petersburg in 1985 - 1988, showed their identity in the quantity of patterns and their molecular weight (7 patterns with molecular weight 14,9; 12,4; 11,8; 10,6; 4,0; 2,8; 12,0 kb).

3.5 Colicin Production

The ability to produce colicin was discovered among 10 strains (0.98%) isolated from patients in 1986 - 1989. All 10 strains produced group I colicin.

3.6 Outer Membrane Proteins

The spectres of the outer membrane proteins had from 10 to 12 patterns. 15 out of 17 strains had 11 identical patterns, and their molecular weights composed the line from 103 to 18kD (103, 88, 77, 71, 58, 53, 48, 40, 35, 33, 18). There was an insignificant difference between the antibiotic sensitive strains. One had fraction 42MD, another fraction 77MD, expressed faintly. As a whole, the spectres of the outer membrane proteins were practically identical among all S. enteritidis phage type 1 strains irrespective of the source and year of isolation, antibiotic sensitivity and colicin production.

3.6 Enterotoxin Production

None of 15 strains irrespective of antibiotic resistance produced any enterotoxin (St or LT).

Conclusion

Nowadays in many countries the growth of the cases of salmonellosis, caused by S. enteritidis, is registered due to poultry meat and egg products. In different countries a definite phage type of S. enteritidis dominates.

In England it is phage type 4 [11]; in Russia it has been phage type 1 during the last 8 years.

Wide circulation of S. enteritidis phage type 1 among humans during 7 years didn't bring about significant changes in biological properties and appearance of new biological species within serovar. Cultural, morphological, serological and fermentative properties, as well as the presence of serotypospecific plasmid, phage type and outer membrane proteins were stable.

The changes were only in acquisition of the properties coded by plasmids: antibiotic resistance and colicin production. They didn't change phage typing of the strains.

The discovery of R-plasmid coding multiresistance of S. enteritidis to antibiotics in some strains with serotypospecific plasmid allows making a proposal that polyresistant clones are formed from the trivial, widely spread variant of S. enteritidis.

Tabl. 1 PHAGE TYPE DESTRIBUTION OF S.ENTERITIDIS STRAINS

Phage type	1985	1986	1987	1988	1989	1990	1991	1992	Total	
									No	%
1	95,6	95,1	82,2	90,6	89,6	87,4	89,8	94,3	1167	88,4
2	-	0,5	1,0	0,4	1,3	0,4	-	-	10	0,8
3	-	-	0,3	-	0,6	-	-	-	5	0,4
4	0,4	-	0,6	-	0,6	0,8	1,4	0,7	13	1,0
6	-	-	-	-	1,0	1,4	2,2	1,6	16	1,2
7	0,6	-	1,0	0,4	0,7	2,0	0,1	0,6	7	0,5
10	0,8	-	6,4	0,6	1,2	-	0,7	-	27	2,0
12	0,8	-	1,2	-	-	2,2	-	1,0	8	0,6
17	-	-	4,3	-	0,3	-	0,5	-	10	0,8
Others*	1,8	4,4	3,0	8,0	4,7	5,8	5,3	4,3	57	4,3

* Other types (9, 16, not typable, not characteristic)

Tabl. 2 ANTIBIOTIC RESISTANCE OF S.ENTERITIDIS PHAGE TYPE 1

Markers of resistance	Amount of strains	%
Cm	24	2,1
Str	10	0,9
Km	6	0,5
Ap	6	0,5
Tc	7	0,6
Cm, Km	6	0,5
Str, Ap	6	0,5
Str, Tc	4	0,3
Cm, Str, Km, Tc	12	1,0
Cm, Str, Km, Ap	11	1,0
Cm, Str, Km, Tc, Ap, Gm	5	0,4
With 1 marker	53	4,5
With 2 markers	16	1,4
With 4 - 6 markers	28	2,4
With 1 - 6 markers	97	8,3

Cm - chloramphenicol; Km - kanamycin; Str - streptomycin;

Tc - tetracycline; Ap - ampicillin; Gm - gentamycin

REFERENCE

1. B.L. Cherkassky, S.Sh. Roxhnova, O.A. Khristyuchina and N.K. Solovyeva. "Current Features of the Epidemic Process of Salmonella Infections in the USSR", *J. Microb., Epid. and Immun.*, No. 1, 1991, pp. 32-36 (Russian).

2. S.Sh. Rozhnova, V.A. Kilesso, N.Z. Aleksandrova. "Etiological Structure of Salmonellosis and Serologic Pattern of Salmonella, Isolated on Different Territories of the Country in 1980 - 1982", *J. Microb., Epid. and Immu.*, No. 11, 1985, pp. 60-63 (Russian).

3. L.B. Hasenson, L.A. Kaftyreva, E.V. Voitenkova and M.J. Nesterova. "Epidemiological and Microbiological Date on Salmonella Enteritidis", *Acta Microbiologica Hungarica*, Vol. 40, No. 1, 1992, pp. 39-43.

4. P.R. Edwards and W.H. Ewing. "Identification of Enterobacteriaceae", 3rd Ed. Minneapolis, MV: Burgess Publishing Co., 1972, . 146-208.

5. V.G. Laszlo, E.Sz. Csorian and J. Paszti. "Phage Typing and Epidemiological Significance of Salmonella Enteritidis Strains in Hungary Between 1976 and 1983", *Acta Microbiologica Hungarica*, Vol. 32, No. 4, 1985, pp. 321-340.

6. C.J. Kado and S.T. Lui. "Rapid Procedure for Detection and Isolation of Large and Small Plasmids", *J. Bacteriol.*, Vol. 145, 1981, pp. 1365-1375.

7. R. Fredericq. "Colicins et Colicinogenic", *Ann. Pastcur.*, No. 5, 1964, pp. 7-17.

8. Zh.A. Volchkevich. "Modified Method of Isolation of Outer Membrane of Pseudomonas Aeruginosa", *Biochem. and Biophys. of Microorganisms*, 1987, pp. 96-100 (Russian).

9. J.P. Vartanyan, M.K. Severtsova and O.I. Vvedensky. "Foot Pad Oedema - Evaluation Test of Enterotoxin E. coli Activity", *Ann. Exp. Biol. and Med.*, No. 2, 1978, pp. 150-152 (Russian).

10. A.G. Dean, Y.C. Ching, B.C. Williams et al. "Test for Escherichia coli Enterotoxin Using Infant Mice", *J. Infect. Dis.*, Vol. 125, 1972, pp. 407-411.

11. E. Coyle, S. Palmer, C. Riberio et al. "Salmonella Enteritidis Phage Type 4 Infection: Association with Hens' Eggs", *Lancet*, Vo. 2, No. 8623, 1988, pp. 1295-1296.

A CHROMOSOME MAP OF SHIGELLA FLEXERI

V.G. Petrovskaya[1], V.M. Bondarenko[1,2] and I.A. Nastichkin

In 1982 a paper was published in which the information about the history of genetic investigations carried out on *Shigella* was summarized and the first provisional map of chromosomes of these bacteria presented (Petrovskaya and Lycheva 1982). The arguments confirming the earlier formulated position (Brenner et al 1973) about a homology of genomes of *Shigella* and *Escherichia* and analugous localization of various genes on their chromosomes were underlined. These data gave a foundation for a recommendation even to unite them into one genus (Collwell ct al 1974).

Key words: chromosome map, *Shigella*, genetic characteristics, pathogenicity, virulence determinants, antigens, virulence plasmids.

1 Gamaleya Institute of Epidemiology and Microbiology, Academy of Medical Sciences ul Gamalei 18, Moscow 123098
2 tel (095-193-61-18)

As *Shigella* are mainly interesting as etiological factors of dysentery, a shorter variant of the chromosome map of these bacteria with underlined regions connected with their pathogenicity was presented. The first variant of the map was included into the collection of the chromosome maps of bacteria (Zakharov and Matselyukh 1986) and the second one was presented in the review of Simmons and Romanowska (1987). In 1985-1988 shorter variants of the chromosome maps of *Shigella flexneri* including only locuses coding for virulence determinants or linked to them were published (Kopecko et al 1985; Hale and Formal 1987. 1988).

At present we think it logical to present a supplemented with new data chromosome map of *Shigella flexneri* (Fig. 1).

It is necessary to underline the following principle of mapping of the *S.flexneri* chromosome: (i) determining the order of transfer of various genes using the Hfr strains of *Escherichia coli* K-12 with a different direction of the marker transfer (Luria and Burrows 1957; Falkow et al 1963); (ii) further conjugation studies emploing constructed Hfr strains of *S.flexneri* with marked strains of *E.coli* K-12 used as recipient strains (Schneider and Falkow 1964; Formal et al 1970; Petrovskaya et al 1972; Sansonetti et al 1983). Analysis of these data led Kopecko et al (1985) to conclude that *Shigella* encoded most of the same genes as *E.coli* and the genes order and map position are thought to be for the most markers identical to that of *E.coli*.

All these investigations made it possible to come to determining the chromosome regions of *S.flexneri* connected with their pathogenicity. The principle of these investigations was as follows: (i) matings of *Shigella* with Hfr strains of *E.coli* K-12 with reference to the loss of some properties after a replacement of the regions with known markers (Formal et al 1970; Petrovskaya and Bondarenko 1971, 1977; Hale and Formal 1986); (ii) a transfer from Hfr strains of *S.flexneri* of some pathogenic properties to *E.coli* K-12 (Formal et al 1970); Makletsova and Petrovskaya 1976; Sansonetti et al 1983); (iii) a transduction analysis of linkage of the discovered genes to corresponding markers (Petrovskaya and Bondarenko 1985; Hale and Formal 1987; Lawlor et al 1987); (iv) using methods of insertion mutagenesis, cloning of the genes helped a more correct determination of localization of genes connected with pathogenicity of *Shigella* (Yoshikawa et al 1988; Hale 1991; Okada et al 1991).

All these data allow a determination of the position of the *S.flexneri* gene relative to the position of analogous genes of *E.coli* K-12 (in minutes). So, Hale (1991) in Table 1 showing the linkage of the rfc gene to the his locus defines this locus as hisA to hisY (44 min).

We presume that the chromosome map of *S.flexneri* presented herein with an indication of the concretely studied loci will be useful for further investigation into the genetic control of pathogenicity.

In the chromosome map of *S.flexneri* shown in Fig.1 the gene symbols not connected with pathogenicity have been included inside the circular map. The position on the map of chromosome locuses connected with the virulence of *S.flexneri* is in several cases approximate and the symbols taken out from the circle. For the fist time in the present map we used a differentiated approach to estimate the role of different regions of the *S.flexneri* chromosome in *S.flexneri* virulence. We distinguished between the determinats of pathogenic factors that are absolutely necessary for virulence expression group (A), and "accessory" determinants group (B). Regulatory genes controlling the expression of genes localized on the virulence plasmid of *S.flexneri* are very important and have been included into group (C). And, finally, we conditionally singled out a group of locuses whose mutations led to the attenuation of bacteria via unknown mechanisms (group D). Under natural conditions the mutations in these genes seem to be induced by different mutagenic factors. Mutants of this type obtained experimentally are used as live vaccines.

In the map these A, B, C, D groups of determinants are marked as , , , and , correspondingly. The circular map of *S.flexneri* (Fig. 1) has been supplimented by Table 1 which lists the phenotypic properties determined by appropriate genes.

In the second part of the paper we present some data about the functions of chromosomal genes connected with pathogenicity of all "ranges" (see Table 2) together with our comments.

We hope it is not useless to present in Table 2 the characteristics of special chromosomal determinants of virulence basing on differential approaches to their assessment. Gene symbols are presented not in alphabetical order as they are in Table 1 but just on the principle of the order of their involvement in realization of the separate stages of the infectious process (Hale andd Formal 1988) with their functional differentiation mentioned above (Table 2). In such order we give also our comments.

Regions of *S.flexneri* chromosome related to their pathogenicity

It is known that pathogenicity is a polydeterminant prperty (Petrovskaya 1967) controlled by the genes of chromosome and plasmid localization (Hale 1991, Ketyi 1984, Kopecko et al 1985, Petrovskaya and Bondarenko 1985)

Shigella and *Shigella*-like enteroinvasive *Esherichia coli* (EIEC) are facultative intracellular parasites for which phases of intracellular penetration, multiplication, and intercellular dissemination are especially important (Timakov et al 1980; Sansonetti et al 1982a; Hale and Formal 1987; Bondarenko and Zhalko-Titarenko 1991).

One of the most important scientific achievements of the past decades in the study of genetics of virulence of *Shigella* and EIEC is the discovery of the role in the pathogenicity of these bacteria of specific high molecular weight plasmids of 120 kDa (180 kb) effecting the synthesis of antigen I phase of smooth (S) type and anvasiveness of *S.sonnei* (Kopecko et al 1980; Sansonetti et al 1982a) and plasmid mm 140 kDa (230 kb) visualized and characterized in *S.flexneri* of different serovars, *S.dysenteriae*, *S.boydii*, and EIEC of 0140, 0143, and 0164 serogroups (Sansonetti et al 1982b; Harris et al 1982).

However, in experimental genetic matings of virulent *S.flexneri* with *E.coli* K-12 as a recipient it was shown that for expression of pathogenicity in *E.coli* it is not only necessary to transfer plasmid mm 140 kDa but additionally at least three regions of the *Shigella* chromosome (Sansonetti et al 1983).

Recently a thesis has been formulated about the presence of a virulence regulon as a common system ensuring pathogenic properties (Maurelli and Sansonetti 1988). It was supposed that there exist "hierarchic" distributions of functions of plasmid and chromosome genes connected with virulence of *Shigella*; plasmid genes determine the adhesion, intracellular penetration, intercellular dissemination, while chromosome gene control the properties responsible for multiplication in the intestinal lumen and tissues (Hale and Formal 1986; Pokrovsky et al 1990) and carry important regulatory genes (Hale 1991).

As in the recently published review of Hale (1991) data are presented about the inv plasmids of *Schigella* and their functions, we have omitted this information from our paper. The second part of the paper deals with the supplements and comments to the data on the role of *S.flexneri* chromosome in expression of their pathogenicity described in a condensed form in the review of Hale (1991).

A. The main determinants of virulence

A role of LPS O antigen in the virulence of gram-negative bacteria is well known (Smith 1977). In a series of brilliant investigations of immunochemical structure of *S.flexneri* LPS O antigen carried out by Simmons (1971) and followed by Lindberg et al (1973) and Kenne et al (1978) with the use of the methylation method, it was demonstrated that *S.flexneri* LPS serovars 1-5 are based on the structure of the group-specific antigen 3,4 (even in the absence of its serological expression). This antigen includes polysaccharide chains containing 2-O-acetyl-L-rhamnose and 1,3-N-acetyl-0-glucosamine.

These data were confirmed in genetic experiments in which a possibility was shown of transition of *S.flexneri* serovars 1-5 into the "y" variants as a result of substitution during conjugation with *E.coli* K-12 Hfr strains of the lac-pro region of their chromosome (Luria and Burrows 1957; Falkow et al 1953; Timakov et al 1970, 1980; Petrovskaya et al 1979).

The genetic control of the synthesis of LPS O antigen (factor 3,4)-w the significance of its R-core and side chains have been described in detail by Hale (1991). In our opinion, further investigations are necessary to elucidate the role of antigen 3,4 in the interaction with the intestinal epithelium, which is ruled out by Japanese investigators (Okamura and Nakai 1977).

To-date we still have no explanation as to why after a substitution of the group-specific antigen 3,4 of *S.flexneri* for antigen 0.25 of *E.coli* containing rhamnose in its structure virulence is preserved, whereas in the case of analogous substitution for antigen 0.8 containing mannose virulence is lost (Gemski et al 1972). It is also unclear why the described by Kim and Corwin (1974) and Mirolyubova et al (1977) factors taking part in the penetration of *Shigella* into HeLa cells are functionally active only with the strain of a homologous LPS structure.

Furthermore, a hypothesis is known about the "opposite" localization of ligands and receptors in *Shigella*. It is postulated that a ligand located on the surface of intestinal mucosa epitheliocytes is a "polysaccharide binding" protein (Izhar et al 1982).

The main structure of antigen VI of *S.flexneri* serovar 6 (*S.newcastle*) includes N-acetyl-galactosamine (Simmons and Romanowska 1987) and in contrast to antigen I-V, the gene determining the synthesis of antigen VI as a gene rfb-a3,4 coding for group-specific antigen 3,4 is localized near the his marker (45 min) of the chromosome map (Makletsova and Petrovskaya 1976). It is possible to suppose that its role is analogous to that of antigen 3,4 in the infectious process.

Aerobactin. A role of Fe^{++} ions in multiplication of pathogens *in vivo* is well known (Payne 1989). The chromosomal gene (iucA) of *S.flexneri* coding for the synthesis of a syderophore (ferrum-associated protein, aerobactin) and additionally of OMP mm 76 kDa (uitA) are necessary for full expression of the virulence potential of these bacteria (Hale and Formal 1986; Lawlor and Payne 1984; Griffiths et al 1985; Lawlor et al 1987; Nassif et al 1988). However, preservation by mutants of their ability to multiply inside HeLa cells (Lawlor and Payne 1984) suggests that aerobactin synthesis has less significance for intracellular multiplication of *Shigella* than the ability to synthesize OMP 21 kDa protein (a product of the pcr gene of inv plasmid pSF140. Aerobactin is more important for multiplication of bacteria in tissues that correspond to the concept of Hale and Formal (1986).

It is important that aerobactin synthesis is not temperature dependent, i.e., it is not a part of the virulence regulon related to the vir plasmid (Maurelli and Sansonetti 1988). It is supposed that *Shigella* sp. could have acquired aerobactin via integration of genes from the *E.coli* pColV (Hale 1991).

Superoxide dismutase. Up to now the gene (sodE) determining the synthesis of supeoxide dismutase has not been mapped. However, its important role in comparison with the wild virulent strain of *S.flexneri* and its transductants (sodB gene) obtained from *E.coli* K-12 has been demonstrated (Franzon et al 1990). This Fe-containing enzyme stimulates the oxidative burst in mouse peritoneal macrophages or human polymorphonuclear leucocytes and is considered as a decisive factor in the pathogenesis of shigellosis (Hale 1991).

It is supposed that this gene is switched on under conditions of O-stress inside monocytes or PNL and possibly after a transition from anaerobic conditions in the lumen to the aerobic medium of epithelial cells (Franzon et al 1990).

Enterotoxins. A question with enterotoxins of *Shigella* and EIEC is not clear. There are distinguished Shiga toxin (ST) and Shiga-like toxins (SLT) including Vero toxins of *E.coli* (O'Brian et al 1987; Ketyi 1985). The most active producent of ST, as it is known, is *S.dysenteriae* I (Bartlett et al 1986). At first the ST gene was mapped near the ldc (lysine decarboxylase) gene (Timmis et al 1985). However, later studies revealed that this gene determines the production of only low level of ST (Sekizaki et al 1987). Besides, it was shown that stx- mutants of *S.dysenteriae* I continued to induce a positive keratoconjuctival reaction (Sereny test) and conserved an invasive ability in HeLa cells (Levine et al 1973).

So a decisive role of ST even in *S.dysenteriae* I (Shiga) in the pathogenesis of shigellosis is not clear. Recently the structural genes of ST (stxA and stxB) have been cloned and a full identity (more than 99% homology) to stx genes of the strains of *S.dysenteria* serovar 1 but not serovar 2 of the same species shown (Strockbine et al 1988). Stx gene was not detected on the chromosome of other studied *Shigella* species: the absence in the bacteria of DNA sequences determining the synthesis of Vero toxins in EIEC was demonstrated. As known, in *Escherichia* the synthesis of these toxins is determined by converting phages (Smith et al 1983).

These data gave a reason not to include the stx gene mapped on the *S.dysenteriae* I chromosome into the chromosome map of *S.flexneri.*

Generally accepted is the presence in *S.flexneri*, *S.boydii*, and *S.sonnei* of both ST and the toxins not reacting with ST sera (Keusch and Jacewicz 1977; Bartlett et al 1986). It was suggested that in *S.flexneri* the low level synthesis of ST is coded by the chromosomal locus near the ldc marker. However, according to Strockbine et al (1988) it is doubtful because of the lack of necessary serological data. However, a connection of this region to the ability of *Shigella* to induce accumulation of liquid in ligated rabbit gut loops has been proved by Hale and Formal (1987, 1988).

We agree with Dr.Hale that its localization is not exact. This question requires further investigations.

B. Additional factors enhancing virulence

As such genes we consider the determinants responsible for the synthesis of serovar(type)-specific antigens I, II, III, IV, V and group-specific antigens 6 and 7,8. The antigens are secondary side chains of LPS (Simmons and Romanowska 1987). As it is known, sporadic cases and even outbreaks of shigellosis may be induced by "y" variants of *S.flexneri* deprived of serovar-specific antigens (-:3,4), So, serovar-specific antigens are not absolutely necessary components for full virulence expression of *S.flexneri.* This was confirmed in genetic experiments (Sansonetti et al 1983). However, studies on genetically connected pairs of strains of these bacteria showed that the strains carrying these additional polysaccharide chains in LPS structure are characterized by greater expression resistance to phagosomal enzymes of macrophages, action of antibodies, and greater ability to multiply in tissues (Ketyi 1974; Bondarenko et al 1975).

However, under natural conditions associated with predominant circulation of the strains carrying serovar-specific antigens and the "type-specific" character of the induced immunity, the role of type-specific antigens is rather significant.

Group-specific antigen 7,8 (Ketyi 1974) and possibly factor 6 presenting secondary chains of antigen 3,7 play an analogus role (Simmons and Romanowska 1987).

Catalase (katFG). To not decisive but additional factors of *Shigella* virulence we attributed katF and katG genes determining catalase synthesis. Double katFG mutants obtained by allelic change with analogous genes of *E.coli* K-12 absolutely lost an ability to synthesize catalase which was shown to be more sensitive to the bactericidal action of macrophages. It is supposed that catalase protects a part of bacterial cells fron oxidative stress though its contribution is rather limited (Franzen et al 1990).

Enterobactin (entABCD). Earlier we pointed out a significant role of Fe^{++} cations in multiplication of pathogenic bacteria *in vivo* and the associate role of *Shigella* gene products determining the synthesis of an aerobactin syderophore of the phenolate type. At the same time, it was shown that about 10% of clinical strains of *S.flexneri* are able to synthesize in addition a syderophore of the catechol type - enterobactin (Payne et al 1983; Payne 1987; Schmitt and Payne 1991). It was shown that genes coding the synthesis and transport of enterobactin are also found on the chromosome of Ent^- clinical isolates of *S.flexneri.*

Genetic studies revealed that Ent^- strains produce smaller quantities of the FepA protein localized on the outer membrane of bacteria. This protein serves as a receptor for ferrienterobactin molecules (Schmitt and Payne 1991). Therefore we thought it important to carry out a study of a possible role of enterobactin in multiplication of *S.flexneri in vivo* as it is not clear why despite enterobactin production this system is conserved in *Shigella.* It is necessary to underline that in Ent^+ strains the synthesis of several proteins is derepressed (Payne et al 1983; Schmitt and Payne 1991).

C. Genes regulating expression of *Shigella* virulence plasmid

Recently a conception has been formulated of a "global central system" in bacteria responding to changes in their environment. This system is "utilized" also by pathogenic bacteria: virulence genes of facultative parasites are "switched off" in a period of life of a pathogen in the external nedium and "switched on" in host organisms. There are positive (activators) and negative (repressors) legulators (Maurelli and Sansonetti 1988; Hromocki and Maurelli 1989).

OmpR-envZ, kcpA loci and virR genes act by inducing or repressing (virR) corresponding locuses on the inv plasmid taking an important role in interaction of *Shigella* with epitelium - a decisive phase of the infectious process. These genes have been fully described and analyzed in recent papers of Bernardini et al (1990), Dorman (1990), Lett et al (1989) and, also, in a condensed form stated in the review of Hale (1991). This is why we confine our report to a short mention and corresponding comments only.

OmpR-envZ locus. At the first stage of the infection ingested *Shigella* encounter an increased gradient of osmolarity on the colon of primates. It is supposed that the two-component regulatory system ompR-envZ promotes survival and penetration of *Shigella* into epitheliocytes inducing the ipaABCD loci of invasion of the inv plasmid of *Shigella* (Bernardini et al 1990); Hale 1991) It is important to underline that these genes, as it was shown by isogenic mating, are at the same map position in *E.coli* K-12 (Formal et al 1965; Hale 1991) which confirms the conception of the presence of a common global system of control responding to the changes in the environment of bacteria mentioned above.

kcpA locus. Whereas the ompR-envZ system in *Shigella* is mainly involved in promoting the invasion into epitheliocytes, the kcpA gene which is a positive regulator of the virG (icsA) gene of inv plasmid promotes the intra- and intercellular spread of *Shigella.* This is an obligatory condition for a possible full expression of virulence of the bacteria (makino et al 1986; Lett et al 1989; Hale 1991).

It is suggested that the kcpA locus generates a protein which acts as a posttranscriptional activator favouring the transcription of a 120 kDa inv plasmid (Hale 1991).

The virR gene of *S.flexneri*, contrary to the above mentioned genes, is a repressor of the invasiveness loci ipaABCD in *Shigella* during the transition from the cultivation temperature of 37°C to 30°C and lower temperatures. It is known that Nature is economic and in an external medium facultative parasites do not necessiate a synthesis of pathogenic factors needed to counteract the defense mechanisms of macroorganisms. The virR gene as a chromosomal gene controls the expression of virulence of *Shigella* has been mapped and cloned (Maurelly and Sansonetti 1988). An analogous tó the virR gene of *Shigella* osmZ gene of *E.coli* K-12 has also been identified (Hromockyi and Maurelli 1989). Recently Dorman et al (1990) and Hilton et al (1990) have formulated a position about the role in regulation and expression of bacterial genes (including genes connected with their pathogenicity) of changes in DNA supercoiling as a response to different signals of the external medium, such as osmolarity, temperature, anaerobic conditions and metabolite influences. The existence is supposed of only a single gene of this type designated as osmZ and linked to the galV marker in *E.coli* K-12 and also in *S.flexneri* (virR gene) (Hulton et al 1990). Data on the significance of DNA supercoiling in the osmotic regulation of gene expression in *E.coli* and *Salmonella typhimurium* have also been obtained (Higgins et al 1988)

It is supposed therefore that the product of the osmZ gene is a "histone-like protein H1" which binds to DNA in a relatively nonspecific manner. Mutations in the osmZ gene are highly pleiotropic and influence the DNA topology and/or chromatin structures and in this way play a role in regulation of gene expression. Disturbances in osmZ influence the expression of many chromosomal and plasmid genes, however, in different directions: part of the genes may be repressed, while the other part either derepressed or not changed at all. Gene expression, as it was mentioned earlier, depends on external influences, temperature, osmolarity, etc.

Studies of the mechanism by which the H1 protein may influence DNA supercoiling have been outlined (Hulton et al 1990). From our point of view, while discussing a question about temperature regulation of bacterial genes it is necessary to pay attention to the existence in *E.coli* K-12 besides the osmZ gene controlling the expression of genes after replacement of bacteria from 28-30°C to 37°C, also of the htpR gene on their chromosome (75 min) or plasmids. This gene regulates the expression of different genes at high temperatures (up to 42°C) (Neidhardt et al 1983). It was shown that at such cultivation temperatures significant changes in the synthesis of different proteins may occur: from their repression up to 20-fold augmentation of the synthesis of some of these proteins. Part of these changes are observes at the transcriptional level. Polypeptide F 39,4 which is suggested to be product of the htpR gene is an effector controlling heat induction of the ATP regulon (Heidhardt et al 1983). As some mutant cells quickly die after cultivation at 42°C, it is supposed that induction of the htpR regulon and heat shock protein synthesis confers some degree of resistance to thermal injury (Yamamori and Yura 1982).

A necessity of studies of this system in *Shigella* is evident as it is known that in the acute phase of shigellosis the temperature may rise up to 39-40°C.

ptsHI genes. A hypothesis has been put forward about the participation of a protein-phosphorilating system in regulation of various physiological processes in bacteria (Saier et al 1990). On the basis of data about the regulation in eukaryotes of different cell functions of posttranscriptional modification of proteins by their phosphorylation, such a system has been studied in prokaryotes. It was shown that in prokaryotes there exist three (but not one) types of protein-phosphorylating systems, namely: classical ATP-dependent protein kinases analogous to the system of eukaryotes and two additional ones (so far unobserved in eukaryotes); a sensor-kinase response regulator and a multicomponent phosphoenolpyruvate-dependent phosphotransferase system (PTS) (Saier et al 1990).

All of the three protein-phosphorylating systems in procaryotes interact, promoting regulation of many physiological factors, such as chemotaxis, nitrogen fixation, modulation, sporulation, induction of transport systems, etc. (Saier et al 1989).

Bacterial protein kinases regulate carbohydrate metabolism, gene transcription, cell differentiation, and such specific processes as phage infection and virulence (Perego et al 1989; Saier et al 1990). The PTS system is complex but well characterized and regulates in *E.coli* and *S.typhimurium* the transport and phosphorylation of many carbohydrates and polyols coupled to it (Gershanovitch et al 1987) as well as a variety of bacterial properties including virulence (Chin et al 1989; Saier et al 1990).

At present the genetic characteristics of the mechanism whereby the growth on glucose, for instance, causes transcriptional stimulation of the pts operon have been investigated. This regulation is dependent on the transport through glucose-specific permease of PTS. Glucose acts like an external signal via a mechanism of signal transcription (De Rense and Danchin 1991).

All these observations gave us a basis to assign the PTS system of *S.flexneri* to a group of regulatory genes.

Studies of fosfomycin-resistant mutants of *S.typhimurium* revealed the presence of variants with deletions in the PTS genes (Gordaro et al 1976). Therefore, in order to detect PTS$^-$ mutants in *S.flexneri*, their fosfomycin-resistant variants were investigated. These mutations induced a pleiotropic effect followed by disturbances in the synthesis of some carbohydrates and a loss of virulence of *Shigella* (Lycheva et al 1980). In cross recombination studies on *S.flexneri* strains 2a and their mutants, and also on marked strains of *E.coli* carrying plasmid AUF-3 including the parts of *E.coli* genes responsible for the synthesis of ptsH and pstI complexes and linked to the purC marker, the existence of the PTS system in *S.flexneri* and full expression and homology of the ptsI and purC genes analogous to *E.coli* K-12 genes have been shown (Gershanovich et 1980).

It is important that avirulent ptsH mutants preserve an ability in adhesion and even penetration into HeLa cells but not in intracellular multiplication (Lycheva et al 1980).

These data allow us to formulate a speculative position suggesting that the regulatory role of the PTS system may in this case be connected with repression of ipaABCD genes of iv plasmid pSF140 which, however, requires further investigations.

It is also important to investigate the possible connection between the regulatory functions of the PTS system and the changes in DNA supercoiling as it was suggested for regulatory genes ompR-envZ and kcpA and demonstrated for the virR (osmZ) gene (Hulton et al 1990).

Gene linked to glpK. This gene was assigned by Hale (1991) to the group of regulatory chromosomal genes connected with plasmid gene expression. It is known that glycerol kinase is the target for the PTS-mediated inhibitor of glycerol uptake (Saier et al 1990). However, it was shown that the mutants with opaque colonies on which this investigation had been carried out have spontaneous insertions in plasmid DNA rather than in the chromosome (Hale 1991). Further investigations are necessary.

D. Locuses of attenuation mutations

Earlier a loss of virulence by *S.flexneri* mutants resistant to rifampicin (Rif^r) was shown (Osada et al 1973). These observations were confirmed and it was then demonstrated that Rif^r mutants of *S.flexneri* lost virulence and simultaneously the ability to synthesize the RNA-polymerase subunit related to mutation in the rpoB gene located in *E.coli* K-12 on 90 min of their chromosome map. The mutants penetrated into epithelial cells but did not multiply intracellularly. The reasons for the lowered virulence in Rif^r mutants are still unclear as these mutants are characterized by changing generation time and sensitivity to temperature. However, such mutants can now be used as vaccine strains (Linde et al 1990).

It is known that *E.coli* and other bacteria responsible for the synthesis of inducible enzymes need cyclic adenosine 3',5'-monophosphate and a corresponding receptor protein (CRP). As the transport of antibiotics is also under the control of these systems, the resistance to some of them is used for an effective positive selection of cya and crp mutants (Apler and Ames 1978). The crp gene in *E.coli* K-12 has recently been cloned and sequenced (Cossart and Gecquel-Sanzei 1982). As to *S.flexneri*, a possibility of obtaining mutants defective in utilization of some carbon sources among the clones resistant to the antibiotic mecillinam has been shown

(Aono et al 1979; Marakusha and Petrovskaya 1981). This is also typical for crp and cya mutants of *E.coli* K-12 also (Marakusha and Petrovskaya 1981).

A transfer of these mutants from the *E.coli* K-12 crp purC strA region to the virulent strain of *S.flexneri* 2a also leads in some recombinants to disturbances in the synthesis of carbon sources (Marakusha and Petrovskaya 1981). In transduction studies it was shown that the crp gene of *S.flexneri* is localized in the str-spc region as in *E.coli* K-12 and co-transduced with the strA gene (Marakusha and Petrovskaya 1981). Mutations in this gene induced a pleiotropic effect expressed as disturbances in utiliziation of some carbon sources, resistance to mecillinam, and lowered virulence. These mutations diminished the ability of shigellas to penetrate into HeLa cells (Bondarenko and Marakusha 1982). A necessity of elucidation of the mechanism whereby mutations in the crp gene influence the virulence of *Shigella* becomes evident.

rpsL, rpsQ, and neaB. As it was shown in *E.coli* K-12 in the chromosomal region str-spc, about a half of the genes responsible for the synthesis of ribosomal proteins, rRNA operon, and some other genes important for the life of bacteria are localized (Post and Nakamura 1980).

Studies of spc, rpsL, prsQ, and nea genes and their mapping on the *S.flexneri* chromosome have been carried out. Localization of genes responsible for the synthesis of ribosomal proteins S_5, S_{12}, and S_{17} was investigated using mutants resistant to spectinomycin, streptomycin, and neamine, respectively. Localization of the neaB gene has been studied with Nea^r mutants.

In *E.coli* K-12 two locuses related to neamine resistance are distinguished: neaA linked to spcA which codes ribosomal protein S_{17} and neaB linked to strA, the product of which has not yet been identified (Bollen et al 1975). Localization of these genes in *S.flexneri* was determined by the frequency of co-transduction in isogenic matings, using Spc_r and Str_r as selective markers. The following order of the genes analogous to those of *E.coli* K-12 (Bachmann 1990) has been shown: spcA - neaA - strA - neaB (Marakusha and Petrovskaya 1980).

Studies of virulence of the transductants obtained using a keratoconiunctival model have led us to conclude that the mutation in the spc gene does not influence the virulence of *Shigella*. In contrast, mutants resistant to 100 mg/ml of neamine lost virulence in 100% of cases. Mutations in strSA and neaB genes cause a full or partial loss of virulence (Marakusha and Petrovskaya 1980).

The loss of virulence by Str^r mutants of several pathogenic bacteria was demonstrated in our earlier studies (Petrovskaya 1967). However, far from all mutations in the strA gene were able to induce the loss of virulence. So, the transfer from *E.coli* to the strains of *S.flexneri* alleles with different ability to restrict pheno- and genotype suppression of con and mis mutations (Breckenridge and Gorini 1970) and influenced their pathogenic properties in a different degree. The transductants which received A2 and A40 alleles lost their ability to induce keratoconjuctivitis in guinea pigs, while A1 and A60 alleles only decreased this ability (Lycheva and Marakusha 1978).

An explanation of differences in the influence of mutations in the spcA gene an virulence in comparison with neaA (rpsQ) and strA (rpsL) was obtained in experiments with *E.coli* K-12 A312. Mutations in genes rpsQ and rpsL are characterized by an ability to influence translation processes expressed by the restriction of action of the amber suppressor in the *E.coli* K-12 strain A312 in contrast with the mutation in the spcA gene (Marakusha and Petrovskaya 1980). These results are consistent with the observations of Strignini and Gorini (1970) concerning the unequal influence of mutations in *E.coli* K-12 strA gene on the efficiency of the amber suppressor action. The data obtained on *Shigella* in relation to SpcA, NeaA, and StrA mutations have been confirmed by *Francisella tularensis* studies (Kormilitsyna and Marakusha 1983) and other pathogenic bacteria including *Salmonella typhimurium*, *Salmonella dublin*, *Yersinia pseudotuberculosis*, and *Legionella pneumophila* (Marakusha et al 1987).

As for *S.flexneri* mutants resistant to neamine (rpsQ) and Str^r mutants (rpsL) lost virulence it was shown that they preserved an ability to penetrate into HeLa cells but not multiply in them. NeaB mutants were characterized by a lowered ability in penetration (Bondarenko and Marakusha 1982). It is of interest to study the possible influence of these mutations on the virulence genes of plasmid pSF140 of *S.flexneri*.

Conclusions

We presume that the chromosome map of *S.flexneri* presented herein answers the following questions: what do we know about the chromosomes of these bacteria; which regions are absolutely necessary for full expression of their virulence including regulatory genes; which genes are additionally important and at least which mutations may be used for obtaining potential vaccine strains.

This map can be supplemented after pur and aro mutations in the corresponding potential vaccine strains of *S.flexneri* will be mapped (Formal et al 1989; Linde et al 1990). Mutants of this type are presently used as "carrier strains" in the construction of gene-engineering vaccines. Additional DNA fragments responsible for the antigenic specificity and immunogenicity of other strains of pathogenic bacteria are transferred to the genomes of such carrier strains (Formal et al 1989).

Appendix

Immunochemical studies of LPS O antigen of different subserovars (a and b) in *Shigella flexneri* revealed that antigen 3,4 makes the basis for LPS of all subserovars. Type (serovar)-specific antigens I-V and group-specific antigens 7,8 and 6 are a result of glycosylation or acetylation (antigens III and 5) of the main structure (Simmons and Romanowska 1987). Cases of spotaneous changes in serological characteristics of *S.flexneri* strains isolated even from one patient (so-called "change of types") are often observed (Mulczyk 1960; Timakov et al 1970; Petrovskaya and Bondarenko 1971; Lachowich et al 1972).

The history of studies on the genetic control of these antigens can briefly be summarized as follows:

(i) in several genetic investigations a loss of antigens I, II, IV, and V by lac^{+} hybrids of *S.flexneri* obtained in crosses with Hfr strains of *E.coli* K-12 has been demonstrated (Luria and Burrows 1957; Timakov et al 1970; Fetrovskaya and Bondarenko 1971, 1985). In the hybrids under study the absence of corresponding glycosyl chains has been confirmed (Manson et al 1970);

(ii) at the same time, converting phages I, II, IV, and V have been isolated and characterized (Matsui 1958; Iseki and Hamano 1959; Okada et al 1960; Giammanco and Natoli 1969; Timakov et al 1980). The results from a series of genetic experiments allowed a conclusion that lac-pro regions of the *S.flexneri* chromosome are the attachment sites for converting phages, since 96-98% of the lac^{+} hybrids of serovars 1a, 2a, 4a, and 5a lose their serovar-specific antigens and become sensitive to the effect of specific converting phages (Timakov et al 1969, 1970). Phage conversion was carried out by bacteriophages IV and V of the recipient strain of *E.coli* K-12 to which the determinant of *S.flexneri* antigen 3,4 (his^{+}:$a3,4^{+}$) had earlier been transferred (Petrovskaya and Nevskaya 1975). Expression of these antigens in the structure of LPS of the recipient strain served as a basis for a conclusion that the above bacteriophages contain in their genome a DNA sequence necessary for the synthesis of uridine diphosphoglucotransferases involved in the synthesis of serovar-specific antigens IV and V (Petrovskaya et al 1972, 1979; Simmons and Romanowska 1987).

It is very important to note that Lac^{+} hybrids of these converted clones of *E.coli* K-12 lose also the corresponding serovar-specific antigens and simultaneously the immunity to specific converting phages (Petrovskaya and Nevskaya 1975).

The aforementioned prophages I, II, IV, and V show an allelic character of heredity.

Analogous converting bacteriophages have been described for group-specific antigens 7,8 and 6 are also the result of corresponding modification of the main structure of antigen 3,4 (Petrovskaya and Bondarenko 1969; Ketyi 1974; Gemski et al 1975). In the latter case phage conversion is also carried out with the expression of antigen 6 on His$^+$ hybrids of *E.coli* K-12 (his$^+$ a,34$^+$) (Gemski et al 1975) The integration site of phage 7,8 in the chromosome is the same lac-pro region; however, the linkage rate of the latter is only 10% (Petrovskaya and Bondarenko 1969).

It is known that double conversion by phages II-7,8 and V-7,8 is possible (Giammanco 1968; Giammanco and Natoli 1969). A possibility of recombination between two prophages (V and 7,8) with the following localization of the markers: aV-immV-imm7,8-lac is suggested. Converting phage 6 is localized in the same region (32% linkage rate) (Bondarenko and Petrovskaya 1977) and determines antigen 6. It may be supposed that antigen III expression also depends on the converting phage 6, becose antigen 6 is a product of acetylation of component 4 (2-O-acetyl-L-rhamnose), while antigen III - the product of acetylation of antigen 3,4 (2-O-acetyl-L-rhamnose-1,3-N-acetyl-glucosamine) (Simmons and Romanowska 1987).

The results of these studies indicate that whereas subserovars of "a" (1:4; II:3,4; IV:3,4; V:4) and "x" variants (-:7,8) are monolysogenic and lose the corresponding converting phages transferred to the "y" variant (-:3,4), subserovars "b" are bilysogenic and change their structure depending on the loss of either or both bacteriophages (in the latter case with a transition to the "y" variants (Timakov et al 1969).

On the basis of these investigations the classification of *S.flexneri* received its genetic explanation and was corrected (Petrovskaya and Bondarenko 1977; Petrovskaya and Khomenko 1979; Brenner 1987).

This scheme gives a possibility to distinguish the genetic variability from mixed and super (consecutive) infections. (Timakov et al 1980).

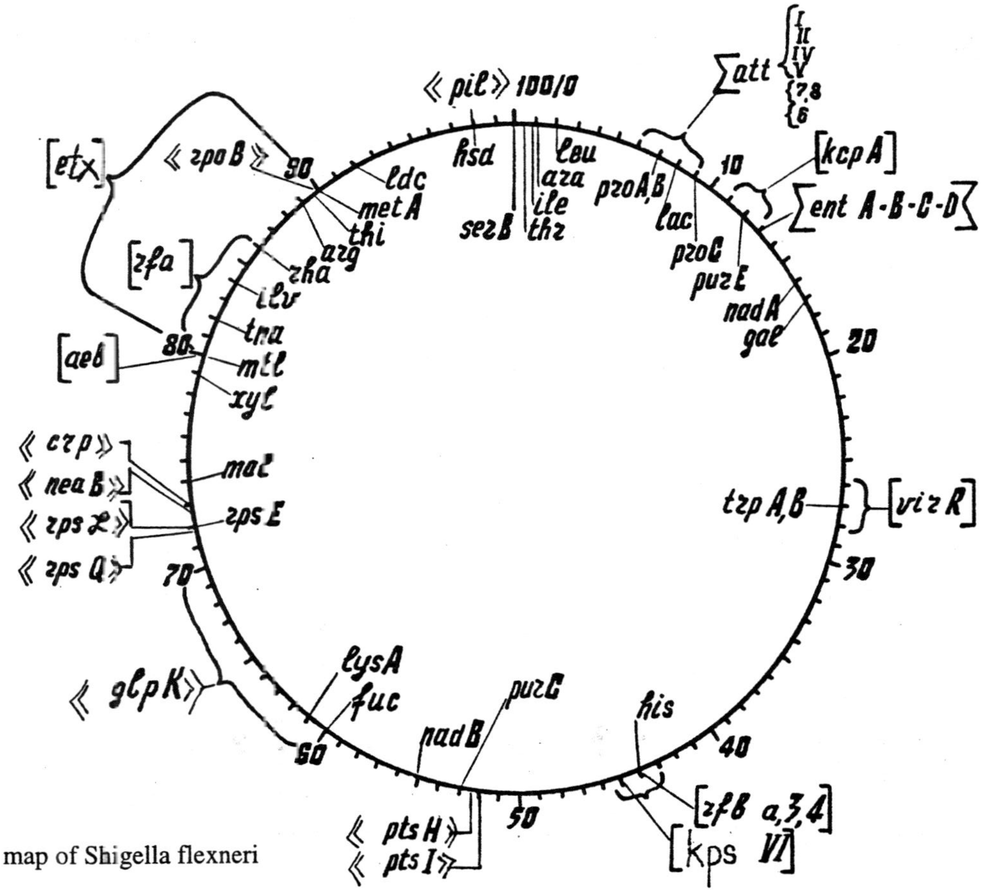

Figure 1. A chromosome map of Shigella flexneri

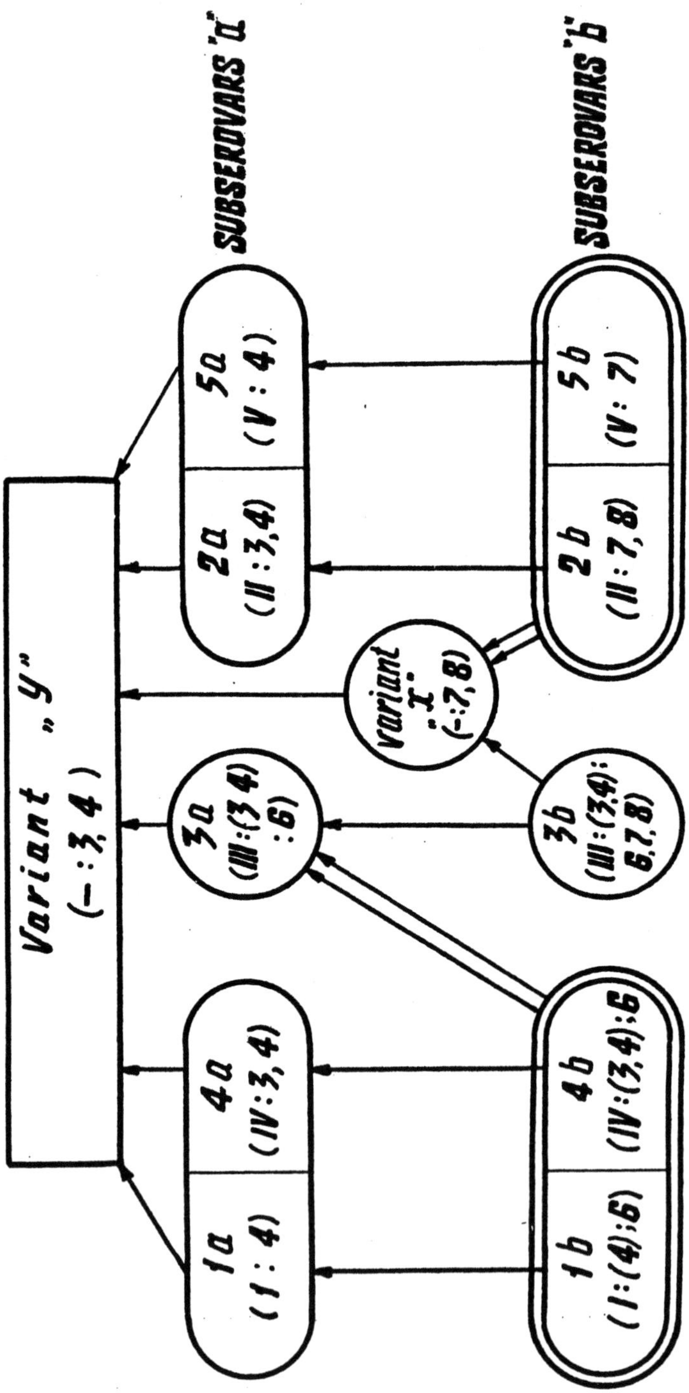

Figure 2. An antigenic variability range of *Shigella flexneri*.

Table 1. GENETIC MARKERS OF SHIGELLA FLEXNERI

Gene symbol	Mnemonic	Map position (min)	Alternative gene symbols, phenotypical traits affected	References
1	2	3	4	5
ara	Arabinose	1	Utilization of arabinose	Falkow et al 1963, Falkow and Formal 1969, Luria and Burrows 1957
arg	Arginine	9o	Synthesis of arginine	Sansonetti et al 1983, Schneider and Falkow 1964, Timmis et al 1985
att I,II, IV,V	Attachment	6-9	Tp-locus; integration sites for correspondent converting phages determining glycosylation of the main structure of LPS (antigen 3,4); conferring serovar (type) specificity (factors I,II,II,IV)	Giammanco and Natoli 1969, Manson et al 1970, Petrovskaya and Bondarenko 1971, Simmons 1971, Simmons and Romanovska 1987, Timakov et al 1969,1970
att 7,8	Attachment	6-9	Integration site for converting phage 7,8 determining glycosylation of antigen 3,4 (group specific factor 7,8)	Giammanco 1968, Petrovskaya and Bondarenko 1969, 1971, 1977
att 6	Attachment	6-9	Possible integration site for converting phage 6 determining acetylation of antigen 4 (group specific factor 6) and expression of antigen III also ?	Bondarenko and Petrovskaya 1977, Gemski et al 1975, Petrovskaya and Bondarenko 1977 Simmons and Romanowska, 1987

1	2	3	4	5
crp	Cyclic AMP receptor protein	74	Resistance to mecillinam	Marakusha and Petrovskaya 1981
ent ABCD	Enterobactin	13	Syderophore of catechol type; only 10% of the strains phenotypically Ent^{+}	Payne 1987,1989, Payne et al 1983, Schmitt and Payne 1988, 1991
etx	Enterotoxin	81-90	flu ; the product induced fluid accumula - tion in rabbit ileal loop	Hale and Formal 1986,1987, Kopecko et al 1985, Sansonetti et al 1983, Timmis et al 1985
fuc	Fucose	60	prd ; utilization of L-fucose	Falkow et al 1963, Formal et al 1971, Gemski et al 1971, Hale and Formal 1988
galE	Galactose	17	Utilization of galactose	Formal et al 1971, Gemski et al 1971, Sansonetti et al 1983
galU	Galactose	28	Glucose-phosphate uridylyltransferase	Dorman et al 1990, Hulton et al 1990
glpK	Glycerolkinase	50-60	Glyerolkinase; corresponds to glpK of E. coli K-12 but is not co-transducted with rha-locus	Kim and Corwin 1974

1	2	3	4	5
his	Histidine	44	Synthesis of histidine	Formal et al 1970, Hale 1991, Petrovskaya and Bondarenko 1971, Schneider et al 1963
hsd	Host specificity	98	Host restriction activity	Falkow and Formal 1969
ileS	Isoleucine	0	Synthesis of isoleucine	Schneider and Falkow 1964
iuc ABCD-iutA		83	Synthesis of aerobactin and 76 kDa iron regulated aerobactin receptor	Griffith et al 1985, Hale and Formal 1986, Hulton et al 1990, Lawlor et al 1987, Lawlor and Payne 1984, Nassif et al 1987
kcpA	Keratoconjunctivitis provocation	12	Positive regulator of virG (icsA) gene of the inv plasmid	Formal et al 1971, Lett et al 1989, Makino et al 1986, Pal et al 1989, Petrovskaya et al 1979, Yamada et al 1989
lac	Lactose	8	Phenotypically lactose negative ; mutations in one of the genes of the lac operon	Giammanco 1966, Timakov et al 1972

1	2	3	4	5
ldc	Lysine decarboxylase	91	Synthesis of lysine decarboxylase	Hale and Formal 1986,1987, Sansonetti et al 1983, Timmis et al 1985
leu	Leucine	2	Synthesis of leucine	Schneider and Falkow 1964
lysA	Lysine	61	Synthesis of lysine	Schneider and Falkow 1964
malA	Maltose	75	Utilization of maltose	Bernardini et al 1990, Formal et al 1965, Hale and Formal 1988, Petrovskaya and Lycheva 1970
mtl	Mannitol	81	Utilization of mannitol	Godard and Hannecart-Pokorni 1975, Hale and Formal 1988, Sansonetti et al 1983
nadA	Nicotinamide adeninedinu-cleotide	17	NicA ; phenotype as a rule is Nad ; Point mutations in one of the genes	Falkow et al 1963, Gemski et al 1971
nadB	Same	56	nicB ; same	Falkow et al 1963, Gemski et al 1971
neaB	Neamine	74	Resistance to neamine	Marakusha and Petrovskaya 1980,1981

1	2	3	4	5
ompR-envZ	Outer membrane protein R ; envelope Z	75	Induction of loci ipaABCD of inv plasmid in response to high osmolarity in colon	Bernardini et al 1989,1990, Formal et al 1965, Hale 1991
pil	Pili	98	fim ; formation of the type 1 pili.	Bondarenko et al 1975a , Falkow et al 1963
proAB	Proline	6	Synthesis of proline	Schneider and Falkow 1964
ptsHI	Phosphoenolpyruvate-dependent phosphotransferase system	52	Resistance to fosfomycine, disturbances in carbohydrate utilization.	Gershanovitch et al 1980, Lycheva et al 1980
purC	Purine	53	ade ; synthesis or purine.	Gershanovitch et al 1980
purE	Purine	12	ade ; synthesis or purine.	Formal et al 1971, Hale and Formal 1987, Lett et al 1989, Pal et al 1989
rfa	Rough	81	LPS core biosynthesis	Godard and Hannecart - Pokorni 1975, Okada et al 1991, Petrovskaya and Lycheva 1970

1	2	3	4	5
rfb a3,4	Rough	44	Synthesis of the main S-specific chain of LPS (group-specific antigen 3,4), containing N-acetylglucosamine	Formal et al 1970, Gemski et al 1972, Okada et al 1991, Petrovskaya and Bondarenko 1971,1977, Sansonetti et al 1983, Simmons and Romanowska 1987
rfb aVI	Rough	45	Synthesis an antigen VI of S.newcastle, containing N-acetylgalactosamine	Makletsova and Petrovskaya 1976, Simmons and Romanowska 1987
rha	Rhamnose	88	Utilization of rhamnose	Falkow et al 1963, Hale and Formal 1987, Petrovskaya and Lycheva 1970
rpoB	RNA polymerase	90	rif; resistance to rifampicin; B-subunit of RNA polymerase	Linde et al 1990
rpsE	Ribosomal protein, small (S_5)	73	strA; resistance to streptomycin	Marakusha and Petrovskaya 1980
rpsL	Ribosomal protein, small (S_{12})	73	strA; resistance to streptomycin	Marakusha and Petrovskaya 1980,1981

1	2	3	4	5
rpsQ	Ribosomal protein, small (S_{17})	73	reaA; resistance to neamine	Marakusha and Petrovskaya 1980,1981
serB	Serine	100	Synthesis of serine	Falkow et al 1963
thr	Threonine	0	Synthesis of threonine	Ketyi 1970, Schneider and Falkow 1964
tna	Tryptophanase	84	ind ; production of indole	Falkow et al 1963, Timakov et al 1969
trp	Tryptophan	28	Synthesis of tryptophan	Falkow et al 1963, Petrovskaya and Lycheva 1970
xyl	Xylose	80	Utilization of xylose	Falkow et al 1963, Petrovskaya and Lycheva 1970
virR	Vir regulator	27	osmZ; repressing of plasmid invasion loci ipaABCD in response to temperature via changing DNA supercoiling	Dorman et al 1990, Hromockyi and Maurelli 1989, Hulton et al 1990, Maurelli and Sansonetti 1988

Table 2. SIGNIFICANCE OF THE CHROMOSOMAL LOCI IN EXPRESSION OF S. flexneri VIRULENCE

Gene symbol	Mnemonic	Linkage	Map position (min)	Phenotype traits affected	Role of the gene products in virulence expression	References
A. Main determinants of virulence						
1	2	3	4	5	6	7
rfa	Rough	mtl	81	Biosynthesis of LPS core	Affect intercellular spread in tissue culture monolayers	Hale 1991, Okada et al 1991, Simmons and Romanowska 1987
rfb a3,4	Rough	his	44	Synthesis of the main specific chain of the LPS O - antigen with N-acetylglucosamine (antigen 3,4)	Protect from chemical factors, local antibodies, and phagocytes in intestinal lumen tissues.	Hale and Formal 1987, Kopecko et al 1985, Okada et al 1991, Okamura and Nakaya 1977, Petrovskaya and Bondarenko 1985
rfb aVI	Rough	his	45	Synthesis of the antigen VI of S.newcastle with N-acetylgalactosamine	The same(?)	Makletsova and Petrovskaya 1976, Simmons and Romanowska 1987

1	2	3	4	5	6	7
iucABCD--iutA		tnaA	83	Synthesis of aero - 76.000 aerobactin and aerobactin receptor	Participation in proliferation in tissues	Griffith et al 1985, Hulton et al 1990, Lawlor et al 1987, Nassif et al 1987, Payne 1989
sodB	Superoxide dismutase	unknown	unknown	Synthesis of Fe-containing superoxide dismutase FeSOD (SodB).	Helps combat oxidative burst of phagocytes.	Franzon et al 1990
etx	Enterotoxin (?)	ldc	81-90	flu ; synthesis of ente - rotoxin (?).	Fluid accumulation in ileal loop.	Bartlett et al 1986, Hale and Formal 1987, Sansonetti et al 1983, Yoshikawa et al 1988

B. Additional genes enhancing virulence

1	2	3	4	5	6	7
att I,II, IV,V	Attachment	pro-lac	6-9	Secondary chains of LPS O-antigen; serovar(type)--specific antigens I,II, IV,V.	Enhance resistance to phagocytosis and antibodies.	Bondarenko et al 1975, Ketyi 1974

1	2	3	4	5	6	7
att 7, 8	Attachment	pro-lac	6-9	Secondary chains – group specific antigen 7,8	The same	Ketyi 1974
att 6	Attachment	pro-lac	6-9	Secondary chains – group specific antigen 6	The same(?)	Bondarenko and Petrovskaya 1977, Gemski et.al. 1975, Petrovskaya and Bondarenko 1977,
ent ABCD	Enterobactin		13	Syderophore of catechol type. Only 10% strains express this ability	Participation in tissue proliferation(?)	Schmitt and Payne 1988,1991
katFG	Catalase	unknown	unknown	Synthesis of catalase	Participation in protection from oxydative burst in phagocytes	Franzon et al 1990

C. Genes regulating expression of the Shigella virulence plasmid pSF140

1	2	3	4	5	6	7
ompR - -envZ	Outer membrane protein R ; envelope Z	malA	75	Induction of loci ipaABCD of inv plasmid pSF140 in response to high osmolarity in colon.	Participation in bacterial invasion under natural conditions.	Bernardini et al 1989, 1990

1	2	3	4	5	6	7
kcpA	Keratoconjunctivitis provocation	purE	12	Positive regulator of virG (icsA) gene of inv plasmid pSF140.	Participation in intercellular spread of Shigella.	Lett et al 1989, Makino et al 1986, Yamada et al 1989
virR	Vir regulator	galU	27	osmZ; repressor of plasmid invasive loci ipaABCD in response to temperature via changing DNA supercoiling	Mutants are invasive in tissue culture when grown at 30°C	Dorman et al 1990, Hromockyi and Maurelli 1989, Hulton et al 1990, Maurelli and Sansonetti 1988
ptsHI	Phosphoenolpyruvate-dependent phosphotransferase system	purC	52	resistance to fosfomycine; pleiotropic effect; disturbance in carbohydrate utilization	mutants are able to adhere and proliferate on the surface of hela cells without penetration inside these cells; Ser	Gershanovitch et al 1980, Lycheva et al 1980
glpK	Glycerol - kinase		50-60	Synthesis of glycerolkinase.	Some mutants are characterized by decreased penetration into HeLa cells	Kim and Corwin 1974

D. Loci mutations in which lead to an attenuation

1	2	3	4	5	6	7
crp	Cyclic AMP receptor protein	strA	74	Resistance to mecillinam	Mutants are characterized by decreased penetration into HeLa cells	Marakusha and Mirolyubova 1981, Marakusha and Petrovskaya 1981
rpoB	RNA polymerase		90(?)	Resistance to rifampicin ; synthesis of the β-subunit of RNA polymerase	Mutants increase generation time, become temperature sensitive; penetrate into epithelial cells but do not proliferate	Linde et al 1990, Osada et al 1973
rpsL	Ribosomal protein, small (S_{12})	spcA	73	strA ; resistance to streptomycin	Mutants penetrate into epithelial cells but do not proliferate.	Marakusha and Mirolyubova 1981, Marakusha and Petrovskaya 1980
rpsQ	Ribosomal protein, small (S_{17})	spcA	73	neaA ; resistance to neamine.	The same	Marakusha and Mirolyubova 1981, Marakusha and Petrovskaya 1980

1	2	3	4	5	6	7
neaB	Neamine	strA	74	Resistance to neamine	Mutants are characterized by decreased penetration into HeLa cells and spread	Marakusha and Mirolyubova 1981, Marakusha and Petrovskaya 1980

REFERENCES

Aono R, Yamasaki M, Tamura G, I979 "High and selective resistance to mecillinam in adenylate cyclase-deficient or cyclic adenosine 3,' 5'-monophosphate receptor protein-deficient mutants of Escherichia coli" Journal of Bacteriology I37,2, 839-845

Apler M, Ames B, I978 "Transport of antibiotics and metabolite analogs by system under cyclic AMP control: positive selection of Salmonella typhimurium cya and crp mutants" Journal of Bacteriology I33, I49-I57

Bachmann B, I990 "Linkage map of Escherichia coli K-I2" Microbiological Reviews 54, I30-I97

Bartlett A, Prado D, Cleary T, Pickering L, I986 "Production of Shiga toxin and other cytotoxins by serogroups of Shigella" Journal of Infectious Diseases I54, 996-I002

Bernardini M, Fontaine A, Sansonetti P, I990 "The two component regulatory system OmpR -EnvZ controls virulence of Shigella flexneri" Journal of Bacteriology I72 6274-628I

Bernardini M, Mounier F, D'Hauteville H, Coquis-Rondon M, Sansonetti P, I989 "Identification of icsA, a plasmid locus of Shigella flexneri that governs bacterial intra- and intercellular spread through interaction with F actin" Proceeding of the National Academy of Sciences of the USA 86 3867-387I

Böllen A, Cabezon T, De Wilde M, I975 "Alteration of ribosomal protein S_{I7} by mutation linked to neamine resistance in Escherichia coli: general properties of neaA mutants" Journal of Molecular Biology 99 (4) 795-806

Bondarenko V M, Marakusha B I, I982 "Biological characteristics of Shigella flexneri transductants acquiring various alleles of StrA locus" Zhurnal Mikrobiologii (8) 35-38 (in Russian)

Bondarenko V M, Petrovskaya V G, I977 "Study of genetic control of the synthesis of type III and group-specific 6 antigens of Shigella flexneri" Zhurnal Microbiologii (4) 79-85(in Russian

Bondarenko V M,Zhalko-Tita..nko V P, I99I "Time course of the formation of a focus of infection in the intestine" Zhurnal Mikrobiologii (8) 23-27 (in Russian)

Bondarenko V M, Petrovskaya V G,Maslova T N, I975 "A study of the role of the type-specific antigen in the interaction of Shigella flexneri with the cultures of epithelial cells and macrophages" Zhurnal Mikrobiologii (5) 37-4I (in Russian)

Breckenridge L,Gorini L, I970 "Genetic analysis of streptomycin resistance in Escherichia coli" Genetics 65 (I) 9-25

Brenner D, I987 "Recommendation on recent proposal for the classification of Shigella" International Journal of Systematic Bacteriology 34 87-88

Brenner D, Fanning G,Miklis G,Steigewalt A,I973 "Polynucleotide sequence relatedness among Shigella species" International Journal of Systematic Bacteriology 23 I-7

Chin A,Feldheim D,Saier M, I989 "Alter transcriptional patterns affecting several metabolic pathways in strains of Salmonella typhimurium which overexpress the fructose regulon" Journal of Bacteriology I7I 2424-2434

Collwell R R,Johnson P,Wan L,Lovelace T E,Brenner D, I974 "Numerical taxonomy and deoxyribonucleic acid reassociation in the taxonomy of some gramnegative fermentative bacteria" Internatinal Journal of Systematic Bacteriology 24 422-433

Cossart P,Gecquel-Sanzey B, I982 "Cloning and sequencing of the crp gene of Escherichia coli K-I2" Nucleic Acids Researches IO I363 I378

De Rense H,Danchin A, I99I "Positive regulation of the pts operon of Escherichia coli: genetic evidence for a signal transduction mechanism" Journal of Bacteriology I73 (2) 727-733

Dorman C,Bhriain N,Higgins C, I990 "DNA supercoiling and environmental regulation of virulence gene expression in Shigella flexner Nature 344 (6268) 789-792

Flakow S, Formal S, 1969 "Restriction in genetic crosses between Escherichia coli and Shigella flexneri" Journal of Bacteriology 100 540-548

Falkow S, Schneider H, Baron L, Formal S, I963 "Virulence of Escherichia - Shigella genetic hybdids for the guinea pig" Journal of Bacteriology 86 I25I-I258

Formal S, Gemski P, Baron L, LaBrec E, I970 "Genetic transfer of Shigella flexneri antigen to Escherichia coli K-I2" Infection and Immunity I 279-287

Formal S, Gemski P, Baron L, LaBrec E, I97I "A chromosomal locus which controls the ability of Shigella flexneri to evoke keratoconjuctivitis" Infection and Immunity 3 73-79

Formal S, Hale T, Kapfer C, I989 "Shigella vaccines" Reviews of Infectious Diseases I(Suppl.)(3) 547-55I

Formal S, LaBrec E, Schneider H, Falkow S, I965 "Restoration of virulence to a strain of Shigella flexneri by mating with Escherichia coli" Journal of Bacteriology 80 835-838

Franzon V, Arondel P, Sansonetti P, I990 "Contribution of superoxide dismutase and catalase activities to Shigella flexneri pathogenesis" Infection and Immunity 58 529-535

Gemski P, Formal S, Baron D, I97I "Identification of two widely separated loci conferring nicotinic acid dependence on wild type Shigella flexneri 2a" Infection and Immunity 3 500-505

Gemski P, Koeltzow D, Formal S, I975 "Phage conversion of Shigella flexneri group antigens" Infection and Immunity II 685-688

Gemski P, Sheahan D, Washington O, Formal S, I972 "Virulence of Shigella flexneri hybrids expressing Escherichia coli somatic antigens" Infection and Immunity 6 I04-III

Gershanovitvh VN, Bolshakova TN, Erlagaeva R S, Glesina M L, Dobrynina O Yu, 1987 "Phosphoenolpyruvate-dependent phosphotranspherase system and regulation of metabolism in a bacterial cell" Uspeki Sovremennoi Biologii 103(2) 173-188 (in Russian)

Gershanovitch V N,Umyarov A M,Burd G I,Bolshakova T.N, Lycheva T A, Petrovskaya V G, 1980 "Mutation in Shigella flexneri resultin in the appearance of phosphomycin resistance avirulent forms with disturbances in carbohydrate utilization" Zhurnal Mikrobiologii (II) 83-88 (in Russian)

Giammanco G, 1966 "Comparsa di mutant -D-galactosidasa-positive in alcuni sierotips di Shigella" Bollettino dell'Instituto Sieroterapico Milanese 45 461-465

Giammanco G, 1968 "Conversion lysogenique de characteres de Shigell flexneri (antigen II de type et complex 7,8 de groupe)" Annales de l'Institut Pasteur Microbiologie 114 63-67

Giammanco G,Natoli D, 1969 "Conversioni antigeniques chez Shigella flexneri" Annales de l'Institut Pasteur Microbiologie 117 16-20

Godard C,Hannecart-Pokorni E, 1975 "Localization sur le chromosome de Shigella flexneri F68 d'un gene rfa implique dans la biosynthèse paroi bacterienne" Archives Internationales de Physilogie et de Biochimie 83 967-972

Gordaro C,Melton T,Stratis J,Atagun M,Gladding C,Herman P,Roseman S 1976 "Fosfomycin resistance: selection method for internal and extended deletions of the Phosphoenolpyruvate: sugar phosphotranspherase genes of Salmonella typhimurium" Journal of Bacteriology 128 (2) 785-793

Griffiths E,Stevenson P,Hale T,Formal S, 1985 "Synthesis of aerobactin and a 76000 dalton iron-regulated outer membrane protein by Escherichia coli K-12 - Shigella flexneri hybrids and by enteroinvasive strains of Escherichia coli" Infection and Immunity 49 69-71

Hale T, 1988 "Organization and expression of plasmid virulence genes in Shigella flexneri" In: Bacteria-Host Cell Interaction Ed Horowitz M (Washington DC:American Society for Microbiology) 253-265

Hale T, I99I "Genetic bases of virulence in Shigella species" Bacteriological Reviews 55 (2) 206-224

Hale T,Formal S, I986 "Genetics of virulence in Shigella" Microbial Pathogenesis I 5II-5I8

Hale T, Formal S, I988 "Virulence mechanisms of enteroinvasive pathogens" In: Virulence Mechanisms of Bacterial Pathogens Ed Roth J A (Washington DC: American Society for Microbiology) 6I-69

Harris J R, Wachsmuth I K, Davis B R, Cohen M L, I982 "High-molecular weight plasmid correlated with Escherichia coli enteroinvasiveness" Infection and Immunity 37 I295-I298

Higgins C, Dorman C,Stirling D, Waddell Z,Booth I,May G,Bremer E, I988 "A physiological role for DNA supercoiling in the osmotic regulation of gene expression in Salmonella typhimurium and Escherichia coli" Cell 52 569-584

Hromockyi A E, Maurelli A T, I989 "Identification of an Escherichia coli gene homologous to virR, a regulator of Shigella virulence" Journal of Bacteriology I7I 2879-288I

Hulton C, Seirafi A, Hinton J, Sidebotham J,Waddell Z, Pavitt G, Owen-Hughes T,Spassky A,Buc H,Higgins C, I990 "Histone like protein HI(H-NS),DNA supercoiling and gene expression in bacteria" Cell 63 (3) 63I-642

Iseki S,Hamano S, I959 "Conversion of type antigen IV in Shigella flexneri by bacteriophage" Proceedings of Japan Academy of Sciences 35 407-409

Izhar M, Nuchamowitz J,Mirelman D, I982 "Adherence of Shigella flexneri to guinea pig intestinal cells is mediated by a mucosal adhesin" Infection and Immunity 35 III0-III8

Kenne L, Lindberg B,Peterson K,Katzenellenbogen E,Romanowska E, I978 "Structure studies of Shigella flexneri O antigens" European Journal of Biochemistry 9I 279-284

Ketyi I, 1970 "Isolation of Hfr derivative by the use of Shigella flexneri 4b modified F-factor" Acta Microbiologica Hungarica 17 167-173

Ketyi I, 1974 "Role in virulence of Shigella flexneri antigens derived from lysogenic conversion" Infection and Immunity 9 931-933

Ketyi I, 1984 "Non-toxin virulence factors of bacterial enteric pathogens" Acta Microbiologica Hungarica 31 1-25

Ketyi I, 1985 "Toxins as virulence factors of bacterial enteric pathogens" Acta Microbiologica Hungarica 32 279-304

Keusch G, Jacewicz M, 1977 "The pathogenesis of Shigella diarrhoeae VI. Toxin and antitoxin in Shigella flexneri and Shigella sonnei infections in humans" Journal of Infectious Diseses 135 552-556

Kim R, Corwin L, 1974 "Mutation in Shigella flexneri resulting in loss of ability to penetrate HeLa cells and loss of glycerol kinase activity" Infection and Immunity 9 916-921

Kopecko D, Baron L, Buysse F, 1985 "Genetic determinants of virulence in Shigella and dysenteric strains of Escherichia coli: their involvement in the pathogenesis of dysentery"

Kopecko D, Washington O, Formal S, 1980 "Genetic and physical evidence for plasmid control of Shigella sonnei form I cell surface antigen" Infection and Immunity 29 207-214

Kormilitsyna M I, Marakusha B I, 1983 "Biological characteristics of spectinomycin resistant strains of Tularemia caucative agent" Antibiotiki 6 434-436 (in Russian)

Lachowicz T, Lagowska M, Kolodynski F, 1972 "The antigenic structure of Escherichia coli and Shigella flexneri hybrids" Archives Roumanaine de Pathologie Experimentale et de Microbiologie 31 39-42

Lawlor K, Payne S, 1984 "Aerobactin genes in Shigella spp." Journal of Bacteriology 160 266-272

Lawlor K, Daskaleros P, Robinson R,Payne S, 1987 "Virulence of iron trasport mutants of Shigella flexneri and utilization of host iron compounds" Infection and Immunity 55 594-599

Lett M,Sasakawa C, Okada N, Yamada M, Komatsu K,Yoshikawa M, 1989 "VirG, a plasmid-coded virulence gene of Shigella flexneri. Identification of the VirG protein and determination of the complete coding sequence" Journal of Bacteriology 171 353-359

Levine M, DuPont N,Formal S, Hornick A,Takeuchi A,Gangarosa E, Snyder M,Libonati F, 1973 "Pathogenesis of Shigella dysenteriae I(Shiga) dysentery" Journal of Infectious Diseases 127 261-270

Lycheva T A, 1976 "Change of virulence of Shigella flexneri after transfer from Escherichia coli K-12 and R mutants of Shigella some fragments of the chromosome correlated with the synthesis of LPS O-antigen" Vestnik Akademii Meditsinskikh Nauk SSSR (I) 34-40 (in Russian)

Lycheva T A, Bondarenko V M,Burd G I,Umyarov A M,Bolshakova T N, Petrovskaya V G,Gershanovich V N, 1980 "Characteristics of attenuated and immunogenic Shigella flexneri mutants selected among phosphomycin-resistant mutants" Zhurnal Microbiologii (6) 58-63 (in Russian)

Lycheva T A, Marakusha B I, 1978 "The influence of transmission of different types of mutations in the strA locus on the virulence of Shigella flexneri" Zhurnal Microbiologii (5) 26-31 (in Russian)

Lindberg B, Lönngren F,Ruden U,Simmons D, 1973 "An investigation of Shigella flexneri O antigens by methylation analysis" European Journal of Biochemistry 32 15-18

Linde K,Dentchev V,Bondarenko V,Marinova S,Randhagen B,Bratoyeva M, Tsvetanov I,Romanova Yu, 1990 "Live Shigella flexneri 2a and

Shigella sonnei I vaccine candidate strains with retained invasiveness but reduced intracellular multiplication" Vaccine 6 25-29

Luria S, Burrows F, I957 "Hybridization between Escherichia coli and Shigella" Journal of Bacteriology 74 46I-468

Makino S, Sasakawa C, Kamata K, Kurata T, Yoshikawa M, I986 "A genetic determinant required for continuous reinfection of adjacent cells on large plasmid in Shigella flexneri 2a" Cell 46 55I-555

Makletsova I.I, Petrovskaya V G, I976 "Genetic control of the synthesis of VI type-specific antigen of Shigella flexneri 6" Zhurnal Microbiologii (I) 44-48 (in Russian)

Manson N, Simmons D, Petrovskaya V G, I970 "Antigenic changes in Shigella flexneri following conjugation with Escherichia coli HfrC" European Journal of Biochemistry I7 472-478

Marakusha B I, Mirolyubova L V, I98I "Influence of mutations conferring resistance to antibiotics with different mechanisms of action on the virulence of Shigella flexneri" Zhurnal Microbilogii (7) 35-38 (in Russian)

Marakusha B I, Petrovskaya V G, I980 "Mapping of mutations in the genes controlling the synthesis of some ribosomal proteins in Shigella flexneri and study of the influence of these mutations on the virulence of these bacteria" Zhurnal Mikrobiologii (I2) 25-3I (in Russian)

Marakusha B I, Petrovskaya V G, Volkovoi K I, I987 "Genetic characteristics and protective properties of neamine resistant mutants of Salmonella typhimurium and Salmonella dublin of the classes Nea^r Str^s and Nea^r Str^r" Zhurnal Mikrobiologii (4) 3-8 (in Russian)

Matsui S, I958 "Antigenic changes in the Shigella flexneri group by bacteriophages" The Japanese Journal of Microbiology 2 I53-I58

Maurelli A, Sansonetti P,1988 "Identification of a chromosomal ger controlling temperature-regulated expression of Shigella viruler ce" Proceedings of the National Academy of Sciences of the USA 85 2820 - 2824

Mirolyubova L V,Petrovskaya V G,Dyuisalieva R G, 1977 "Comparative characteristics of cell free filtrates of Shigella flexneri of d fferent virulence" Zhurnal Microbiologii (5)29-33 (in Russian)

Mulczyk M, 1960 "Badania nadzmiennoscia antigenowa paleczek Shigel la flexneri 1b" Archivi Immunologii i Terapii doswiadczenie 8(1) 1-36

Nassif X, Mazert M, Mounier J,Sansonetti P, 1987 "Evolution with an iuc::Tn10 mutant of the role of aerobactin production in the virulence of Shigella flexneri" Infection and Immunity 55 1963-1969

Neidhardt F, Van Bogelen R, 1981 "Positive regulatory gene for tem perature-controlled proteins in Escherichia coli" Biochemical ar Biophysical Research Communication 100 894-900

Neidhardt F,Van Bogelen R, Lan E, 1983 "Molecular cloning and expression of a gene that controls the high temperature regulon of Escherichia coli" Journal of Bacteriology 153(2) 597-603

O'Brien A,Newland F,Miller S,Holmes R,Smith H,Formal S,1987 "Shiga like toxin-converting phages from Escherichia coli strains that cause hemorrhagic colitis or infantile diarrhea" Science 226 694-696

Okada M,Hoshino P,Karato T, 1960 "Lysogenic conversion of somatic antigen of Shigella flexneri by bacteriophage" Science(Tokyo)30 90-95

Okada N, Sasakawa C,Tobe T,Yamada M,Nagai S,Talukder K,Komatsu K, Kanegasaki S,Yoshikawa M, 1991 "Virulence-associated chromosomal loci of Shigella flexneri identified by random Tn5 insertion mutagenesis" Molecular Microbiology 5 187-195

Okamura N,Nakaya R, 1977 "Rough mutant of Shigella flexneri 2a that penetrates tissue culture cells but does not evoke kerato-conjunctivitis in guinea pigs" Infection and Immunity 17 4-8

Osada Y, Une T,Nakayo M,Ogawa H, 1973 "Virulence of rifampicin resistant mutants of Shigella flexneri and enteropathogenic Escherichia coli with special reference to their cell invasiveness" Japanese Journal of Microbiology 17 243-249

Pal T,Newland F,Tall B,Farmal S,Hale T,1989 "Intracellular spread of Shigella flexneri associated with the kcpA locus and a 140-kilodalton protein" Infection and Immunity 57 477-486

Payne S, 1987 "Iron trasport in Shigella and Vibrio species" In: Iron Transport in Microbes,Plant and Animals Eds Winkelmann G, Van Der Heln D,Neilands J, New York:VCH Publishers 99-109

Payne S,1989 "Iron and virulence" Molecular Microbiology 3 1301-1306

Payne S, Niessel D,Peixotta S,Lawlor K, 1983 "Expression of hydroxamate and phenolate syderophores by Shigella flexneri" Journal of Bacteriology 155 949-955

Perego M,Cole S,Burbulys D,Trach K,Hoch J,1989 "Characterization of the gene for a proteine kinase which phosphorylates the sporulation-regulatory proteins Spo OA and Spo OF of Bacillus subtilis" Journal of Bacteriology 171(11) 6187-6196

Petrovakaya V G, 1967 "The problem of bacterial virulence" Leningrad: Meditzina 263p (in Russian)

Petrovskaya V G, Bondarenko V M,1969 "Phage control of the group-specific antigen 7,8 synthesis in Shigella flexneri" Vestnik Akademii Meditsinskikh Nauk SSSR (11) 6-15(in Russian)

Petrovskaya V G, Bondarenko V M,1971 "Genetic control of the Shigella flexneri type and group antigens" In: Recent Advances in Microbiology Eds Perez-Miravete A, Pelarz D, Mexico City 389-398

Petrovskaya V G, Bondarenko V M, 1977 "Recommended corrections to the classification of Shigella flexneri on a genetic basis" International Journal of Systematic Bacteriology 27 171-181

Petrovskaya V G, Bondarenko V M, 1985 "Genetic basis of virulence of dysentery bacteria" Vestnik Akademii Meditsinskikh Nauk SSSR (10) 7-13 (in Russian)

Petrovskaya V G, Bondarenko V M,Mirolyubova L V, 1979 "The problem of interaction of Shigella with epithelial cells" Zentralblatt fur Bakteriologie Microbiologie und Hygiene Series A 243 57-62

Petrovskaya V G, Khomenko N S, 1979 "Proposal for improving the classification of members of the genus Shigella" International Journal of Systematic Bacteriology 20 400-402

Petrovskaya V G, Lycheva T A, 1970 "Keratoconjunctivitis test in Shigella flexneri 2a hybrids obtained in cross with different strains of Escherichia coli K-12" Annales de l'Institut Pasteur Microbiologie 118 761-766

Petrovskaya V G, Nevskaya N A, Khomenko N A, 1972 "Reproduction of phage conversion of the somatic antigen of Shigella flexneri on hybrids of Escherichia coli" Genetika 8 (9) 115-118 (in Russian)

Pokrovskii V I, Bondarenko V M,Polotskii Yu E, Yushchuk N D, 1990 "Modern views on the pathogenesis of dysentery: the role of invasion and toxic action of Shigella" Zhurnal Mikrobiologii (5) 101-110 (in Russian)

Post L, Nakamura M, 1980 "DNA sequences from the str operon of Escherichia coli" Journal of Biological Chemistry 255 4660-4666

Saier M, 1989 "Protein phosphorylation and allosteric control of inducer exclusion and catabolite repression by the bacterial phosphoenolpyruvate: sugar phosphotranspherase system" Microbiological Reviews 53 109-120

Saier M. Wu L, Reizer J, 1990 "Regulation of Bacterial physiologic process by three types of protein phosphorylating systems" Elsevier Science Publishers Ltd (UK) 391-395

Sansonetti P, Hale T, Dammin G, Kapfer C,Collins H, Formal S,1983 "Alteration in the pathogenicity of Escherichia coli K-12 after transfer of plasmid and chromosomal genes from Shigella flexneri" Infection and Immunity 39 1392-1402

Sansonetti P,Kopecko D, Formal S, 1982a "Shigella sonnei plasmids: evidence that a large plasmid is necessary for virulence" Infection and Immunity 34 75-83

Sansonetti P, Kopecko D, Formal S, 1982b "Demonstration of the involvement of a plasmid in the invasive ability of Shigella flexneri" Infection and Immunity 35 852-860

Schmitt M, Payne S, 1988 "Genetics and regulation of enterobactin genes in Shigella flexneri" Journal of Bacteriology 170 5579-5587

Schmitt M, Payne S, 1991 "Genetic analysis of the enterobactin gen cluster of Shigella flexneri" Journal of Bacteriology 173 (2) 816-825

Schneider H,Falkow S, 1964 "Characterization of a Hfr strain of Shigella flexneri" Journal of Bacteriology 88 682-689

Schneider H, Falkow S, Baron L, 1963 "Genetic homology between Shigella flexneri 2a and Escherichia coli " Genetics 48 907-911

Sekizaki T, Harayama S,Brazil G,Timmis K, 1987 "Localization of stx, a determinant essential for the high-level production of Shiga toxin by Shigella dysenteriae serotype 1 near pyrF and generation of stx transposon mutants" Infection and Immunity 55 2208-2214

Simmons D, 1971 "Immunochemistry of Shigella flexneri O antigen: a study of structure and genetic aspects of the biosynthesis of cell-surface antigens" Bacteriological Reviews 35 117-123

Simmons D, Romanowska E, 1987 "Structure and biology of Shigella flexneri O antigens" Journal of Medical Microbiology 23 289-302

Smith H, 1977 "Microbial surfaces in relation to pathogenicity" Bacteriological Reviews 41 475-488

Smith H, Green P, Parsell Z, 1983 "Vero cell toxin in Escherichia coli and related bacteria: transfer by phage and conjugation a toxic action in laboratory animals, chickens and pigs" Journal of General Microbiology 129 3121-3137

Stocker B, 1988 "Auxotrophic Salmonella typhi as live vaccine" Vaccine 6 (2) 141-145

Strignini P, Gorini L, 1970 "Ribosomes mutations affecting efficiency of amber-suppression" Journal of Molecular Biology 47 517-530

Strockbine H, Jackson M, Sang L,Holmes R,O'Brien A, 1988 "Cloning and sequencing of the genes for Shiga toxin from Shigella dysenteriae type 1" Journal of Bacteriology 170 1116-1122

Timakov V D, Petrovskaya V G, Bondarenko V M, 1969 "A study of genetic control of type-specific antigen synthesis in the subgroup B Shigellae" Vestnik Akademii Meditsinskikh Nauk SSSR (11) 48-58 (in Russian)

Timakov V D, Petrovskaya V G, Bondarenko V M, 1970 "Studies on the genetic control of Shigella subgroup B type specific antigens Annales de l'Institut Pasteur Microbiologie 118 1-9

Timakov V D, Petrovskaya V G, Bondarenko V M, 1972 "Genetic data concerning Shigella flexneri serotype 5 and 6" International Journal of Systematic Bacteriology 22 149-154

Timakov V D, Petrovskaya V G, Bondarenko V M, 1980 "Biological and Genetical Characteristics of Shigella bacteria" Moscow: Meditsina 269p (in Russian)

Timmis K, Clayton C, Sekizaki T, 1985 "Localization of Shiga toxin gene in the region of Shigella dysenteriae 1 specifying virulence functions" FEMS Microbiology Letters 30 301-305

WHO: "Research priorities for diarrhoeal diseases vaccines. Memorandum from WHO mee ting" Bulletin of the WHO 1991 69(6) 667-676

Yamada M, Sasakawa C, Okada N, Makino S, Yoshikawa M, 1989 "Molecular cloning and characterization of chromosomal virulence regio kcpA of Shigella flexneri" Molecular Microbiology 3 207-213

Yamamori T, Yura T, 1982 "Genetic control of heat-shock protein synthesis and its bearing on growth and thermal resistance in Escherichia coli K-12" Proceedings of the National Academy of Science of the USA 79 860-864

the USA 79 860-864

Yoshikawa M, Sasakawa C, Makino S, Okada N, Lett M, Sakai T, Yamada M, Komatsu K, Kamata K, Kurata T, Sata T, 1988 "Molecular genetic approches to the pathogenesis of bacillary dysentery" Microbiological Sciences 5 (11) 333-343

Zakharov I A, Matselyukh B P, 1986 "Genetic Maps of Microorganisms" Kiev: Naukova Dumka (in Russian)

Release of Protein from Escherichia Coli as Universal Bacteria Response to Stress

E.K. Roshchina and L.N. Petrov

Summary

Fluorescence techniques were used for evaluation of protein concentration. Tryptophan-containing protein release was detected as Escherichia coli M-17 response to varied stress effects. Kinetics of protein release from E. coli cells during the storage in physiological solution at room temperature and incubation of cell suspensions at 42, 48, and 55 C are given in full detail below. The kinetic curves of protein accumulation were described by C,T-parameters, where C is the steady-state concentration of extracellular proteins and T is the time until this concentration is established. It was found that C did not depend on either intensity or nature of stress effects, whereas T did. Intensive protein release was accompanied by the maintenance of culture viability at the initial level. Cell death, according to exponential low, progressed after the concentration had increased to the steady-state value. The above-mentioned facts permit classifying protein release as universal protective (or adaptive) bacteria response to stress. Kinetic analysis of ultraviolet-absorbing compound release showed that several components absorbing at 280 nm could take part in adaptive processes. There is experimental proof that transport systems may be involved in protein and UV-absorbing compound release.

INTRODUCTION

It has long been known that stress effects on microbial cells bring into existence a large number of cell components on the surrounding medium among which are proteins, amino acids, nucleic acids, lipopolysaccharides, etc. It is widely believed that metabolite escape from cells is due to the serious injury of the cell envelope or even to the lysis in some parts of the cell population. Today, however, the literature in which the appearance of extracellular compounds is considered from the opposite viewpoint, namely, as protective cell response, is available [1,2,3,4]. Melekhov [1,2] has formulated the principle of damage regulation and protective cell response on the basis of numerous experimental data. According to which the gist of the protective response is retardation in cell metabolism. The author believes that one way of metabolism intensity decrease can be realized through different cell compound release (potassium ions, for example). In general, it can be suggested that two types of compounds are in the extracellular fluid: some of these compounds are the result of cell envelope disruption and others are involved in the regulatory process dedicated to microorganism survival under stress conditions. A lot of publications devoted to the investigation of the nature and functions of so-called stress proteins led us to the idea that extracellular proteins (similar to intracellular ones) accomplish the protective function.

In the present paper we tried to justify experimentally the hypotheses for the protective function of proteins released from E. coli under unfavorable conditions, such as storage and heat shock.

MATERIALS AND METHODS

ORGANISM

Escherichia coli M-17, a laboratory stock culture previously obtained from Tarasevich State Kontrol Institute (Moscow), was used in all experiments.

MEDIA

The glucose-mineral medium was used as a growth medium in cell production for storage, heat treatment, etc. The medium composition was the following (g/l): glucose - 3.0, $Na_2HPO_4*2H_2O$ - 1.0, KH_2PO_4 - 1.0, ammonium citrate - 4.5, NaCL - 5.0, Na_2CO_3 - 1.5, $MgSO_4 * 7H_2$ - 2.0, nicotinic acid - 0.005, pH 7.4.

Endo Agar was used as plating medium to estimate the number of viable cells.

CULTURE PREPARATION

Cells were grown into stationary phase at 37 C shaking. These were centrifuged, washed once with 0.85% NaCl (physiological solution) and resuspended in fresh 0.85% NaCl to the concentration of about $5*10^9$ cells/ml with optical density at 460 nm).

STORAGE AND HEAT TREATMENT

Cell suspensions were stored in glass flasks with cotton-gauze plugs at room temperature (20 + 2) C with 14 day intervals.

For heat treatment 2-ml cell suspension portions were transferred to a series of tubes. The tubes were placed in a temperature-controlled water bath in tandem at regular intervals. When the experiment was over all the tubes were taken out of the water bath at a time. Incubation temperature was 42, 48 and 55 C. In other experiments it varied from 22 to 100 C and the cell suspensions were kept at each temperature for 15 minutes.

After storage and heat treatment cells were removed by centrifugation at 3,000 g for 10 min. The supernatant fluid from these cells was diluted (1:5) and analyzed.

MEASUREMENTS OF LEAKAGE PROTEINS FROM CELLS

Measurements of proteins in supernatant fluids were made in Hitachi model 650-60 fluorescence spectrophotometer when exitation was at 280 nm. All the obtained emission spectra had a maximum at 330-340 nm. This emission band is typical of tryptophan residues incorporated into proteins. Tryptophan, in free solution, shows λ_{max} of emission at about 350 nm ([5] and our data). The fluorescence intensity at λ_{ex} = 280 nm and λ_{em} = 340 nm was used at the measure of tryptophan-containing protein concentration since this parameter varies linearly in different concentrations, if optical density is low enough at 280 nm.

RESULTS AND DISCUSSION

Relying on our experiments, we believe that protein release from E. coli under unfavorable conditions could be considered as universal cell response to stress factors. The protein accumulation in supernatant fluid was observed when the bacteria were transferred from the culture medium to distilled a water or physiological solution, kept in physiological solution at room temperature for some days, heat treated, exposed to electromagnetic radiation, lyophilized, etc. E coli cell behavior during the storage in physiological solution at room temperature and heating from 42 to 55 C were studied in considerable detail.

Fig. 1 illustrates protein appearance kinetics in the extracellular fluid during E. coli storage in physiological solution at room temperature and shows that the rate of the protein accumulation is maximum for the first two days of the cell storage and then it reduces to zero gradually. The intensive release of proteins from E. coli leaves cell viability unchanged. On the contrary, the inability of cells to liberate proteins is accompanied by cell death. The kinetics of the other component release from E. coli is also plotted in this figure. This component can be assigned to peroxidation products of a minor fraction in the bacterial lipids. Its fluorescence lies in the blue range and the correlation coefficient between fluorescence intensity emitted by the component and the number of viable cells has a negative sign and differs reliably from zero [6]. The comparison of curves 1 and 2 (Fig. 2) shows that the destructive oxidation process does not take place rapidly until cells lose their capacity to release proteins. In other words, as long as protein release occurs, cell viability is kept constant since destructive processes, which lead to cell death, cannot take place. These findings allow classifying protein release as E coli defence reaction during the storage.

Fig. 2 demonstrates protein appearance kinetics in the extracellular fluid during the incubation of E coli suspensions at 48 C. The choice of temperature derived from the fact that cells could leave their viability unchanged for some time (30-45 min), being important for the studies on defence reactions. As with storage (Fig. 1), the rate of protein accumulation is the greatest at the beginning of the incubation and then it gradually reduces. In contrast to the preceding experiment, the rate of protein accumulation does not go down to zero during culture incubation at 48 C, although this can be sometimes realized (Fig. 4). This is not surprising, since at about 50 C the considerable realignment in the structure of E coli M-17 outer membrane can be detected, using N-phenyl-1-naphtylamin (1-NPN) as a fluorescent probe [7]. An insignificant accumulation of proteins during the prolonged incubation is due to cell envelope destruction and unrelated to the processes under investigation. The curves for viability and protein release have inflection point on the same spot of the time axis. An increase in the rate of culture death in an answer to a decrease in that of protein release from bacteria exposed to high temperature allow classifying extracellular proteins as compounds participating in defensive processes.

It was found that curve 3 on Fig. 2 had no inflection point, meaning that the initial reason of cell death is due to the realignment of E. coli outer membrane when heated, rather than to lipid peroxidation processes.

It is worth noting that according to exponential law the number of viable cells in culture decreases after the release of proteins stops. By this law, the death probability is constant at each instant of time. The low is realized under sterilization conditions when stress effect is so great that no adaptation is possible. Hence, exponential law of cell death can testify, in our case, that adaptation processes are over.

Since protein release from bacteria is accompanied by the maintenance of cell viability at an initial level under different effects such as storage (or starvation) and heat shock, this is suggested to be classified as universal protective (or adaptive) response of bacteria to stress.

Fig. 3A, 3B present the kinetic curves of protein release from E. coli during the incubation of cell suspension at 55, 48 C (heat shock) and 42 C. For the description of these curves two parameters may be introduced: the steady-state concentration of extracellular proteins (C) and the time until this concentration is established (T). It is worth noting that the incubation of cell suspension at 42 C should be defined as storage or starvation (not heat shock) since the bacteria retain the capability for propagation under these conditions. The curve analysis presented in Fig. 3 shows that the steady-state concentration in extracellular proteins depends on neither the heat shock temperature nor the nature of stress effect. The differences in character or intensity of stress effect influence only T-parameter. Thus, the steady-state concentration of extracellular proteins is established at 55 and 48 C for some minutes, at 42 C for some hours and at 22 C for some days (Fig. 1).

Ultraviolet-absorbing material release from E. coli was also detected in our experiments because it is still the most widespread parameter of the heat-treated bacteria behavior. Fig. 4 gives the kinetic curves of compound release absorbing at 280 nm from heat-treated E. coli cells. As before, these kinetic curves may be described by C,T-parameters, except the ones at 55 C. As was mentioned above, the considerable rearrangement of E. coli cell envelope takes place at about 50 C. This is just the condition under which uncontrolled release of UV-absorbing material from cells can be realized. The steady-state concentration of 208 nm absorbing material in extracellular fluid is a temperature-dependent magnitude, as opposed to the data for proteins. The higher the temperature of incubation gets, the more the steady-state concentration of 280 nm absorbing compounds is.

In our opinion, the experimental data for proteins (C does not depend on temperature) do not contradict to ones for 280 nm absorbing compounds (C depends on temperature). The matter is that we deal with one type of components (proteins) in the former case and with a large set of different components (proteins, nucleic acids, vitamins, co-enzymes, etc.) in the latter one [8]. If the steady-state concentration of each component is defined by its nature and does not depend on a stress type, as was mentioned above for proteins, the temperature dependence of the steady-state concentration for 280 nm absorbing compounds may be attributed to some component involvement in the adaptive process. In other words, the number and nature (chemical structure) of the extracellular compounds is dictated by the stress conditions.

The kinetic curves of 260 nm absorbing compounds do not flatten out. This can denote that the concentration of some compounds absorbing at 260 nm increases progressively and does not go into a steady-state level and these compounds are not involved in the adaptive process.

In the experiments presented in Fig. 3,4 the release of the metabolites from E. coli cells was plotted as a time function at the fixed temperature of incubation. In the experiments given in Fig. 5 the incubation time was constant (15 min) and the temperature variable. As shown in Fig. 5, the concentration of all the components increases monotonically, but there is an anomaly at about 60 C. If temperature is less than 55 C, the concentration of proteins increases with a rise in temperature since the time of the steady-state concentration establishment is more than 15 minutes. If temperature is more than 65 C, the protein concentration also increases, probably due to the injury

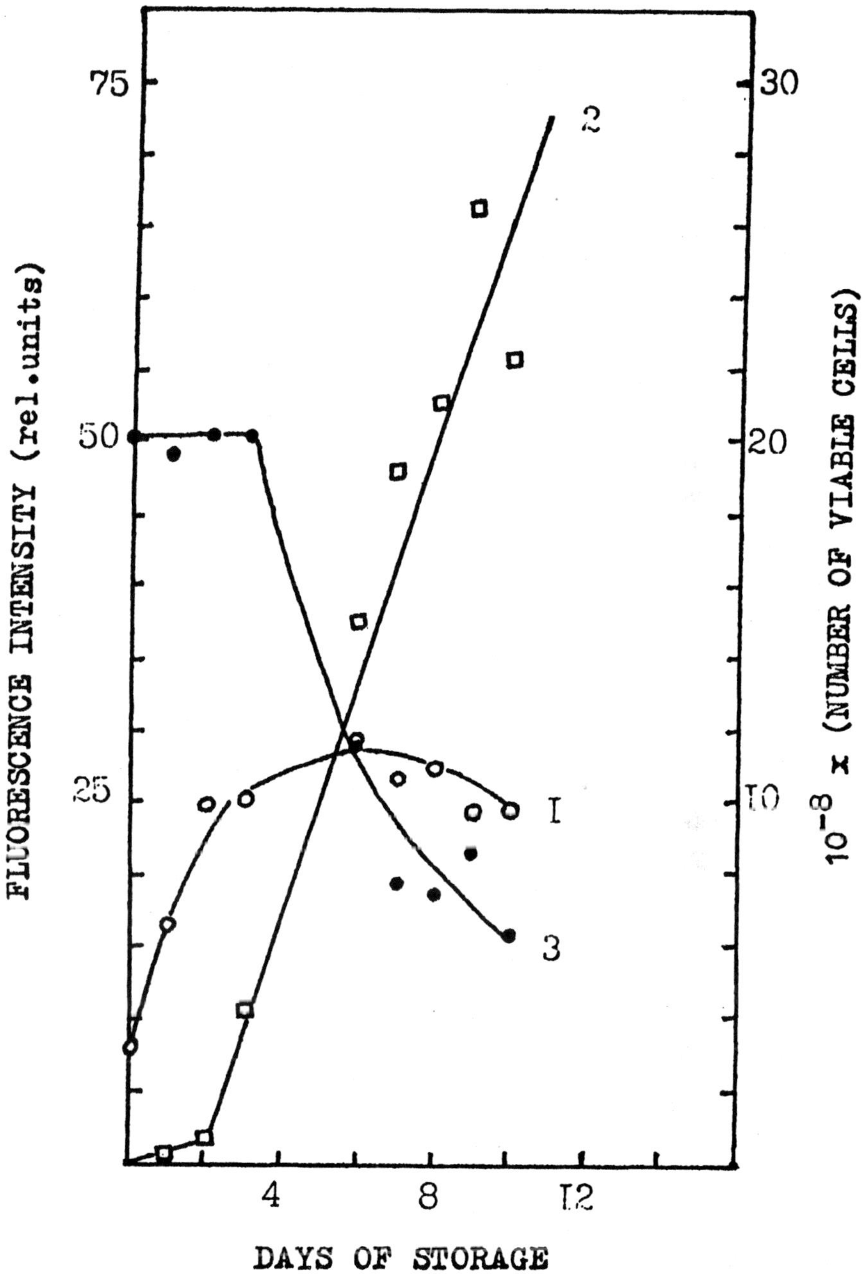

Figure 1 Fluorescent compound release from E. coli M-17 during the storage in physiological solution at room temperature. Fluorescence intensity of supernatant fluid at 1) $\lambda_{ex} = 280$ nm, $\lambda_{em} = 340$ nm (proteins), 2) $\lambda_{ex} = 350$ nm, $\lambda_{em} = 430$ nm (lipid peroxidation products), 3) number of viable cells.

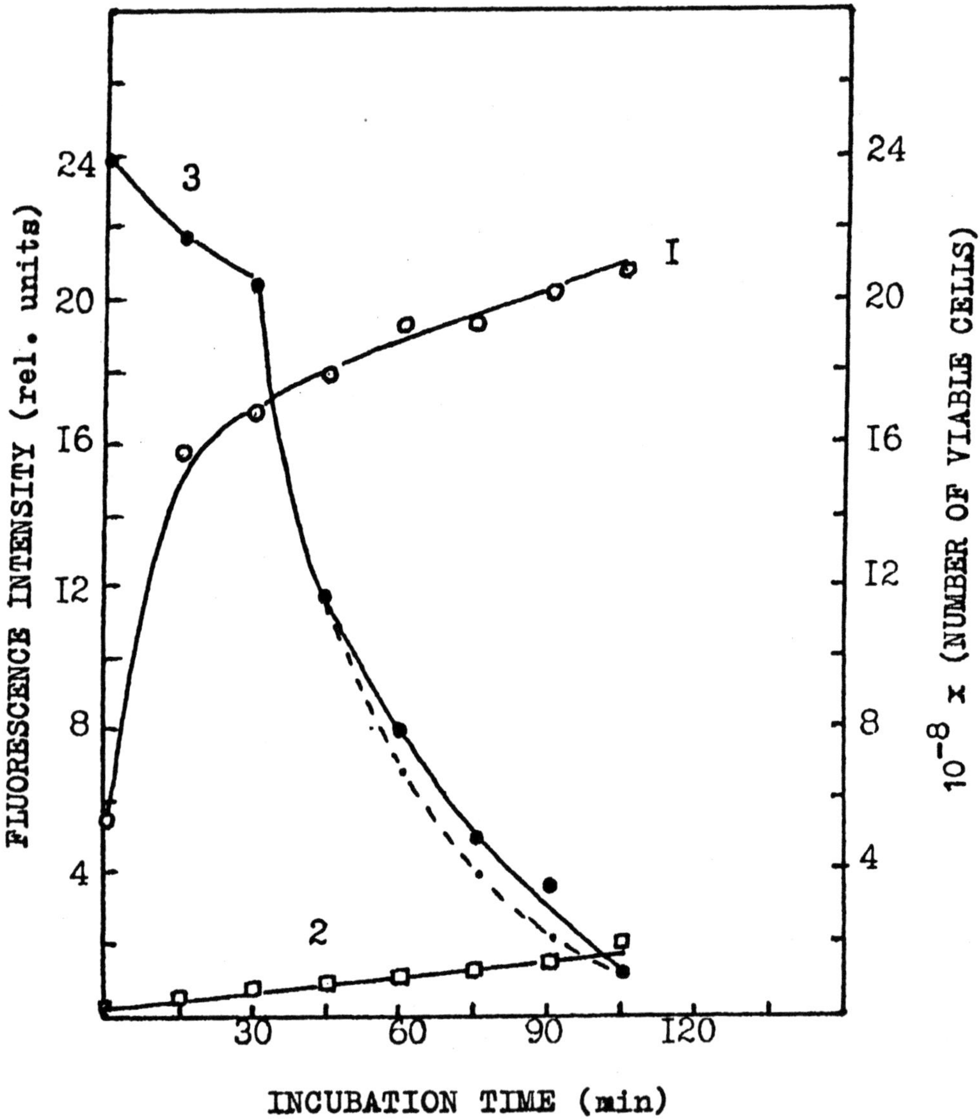

Figure 2 Fluorescent compound release from E. coli M-17 during incubation at 48°C (heat-shock). Fluorescence intensity of supernatant fluid at 1) $\lambda_{ex} = 280$ nm, $\lambda_{em} = 330$ nm (proteins), 2) $\lambda_{ex} = 360$ nm, $\lambda_{em} = 420$ nm (lipid peroxidation products), 3) number of viable cells, experrimental data (continuous line), calculation by exponential low (broken line).

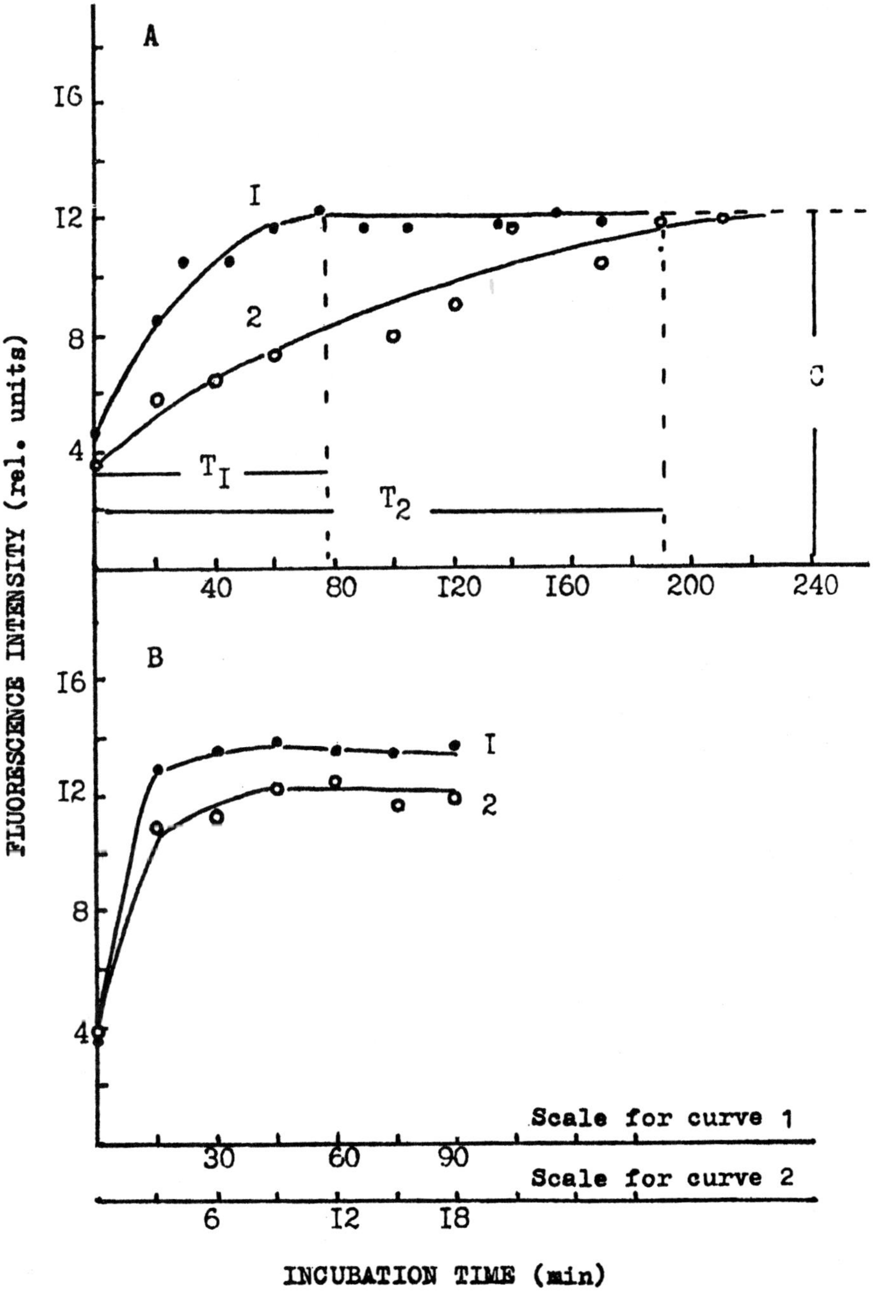

Figure 3 Fluorescent compound release from E. coli M-17 during incubation in physiological solution at A) 48°C (1), 42°C (2), c = 5.2∗10^9 cells/ml; B) 48°C (1), 55°C (2), c = 5.0∗10^9 cells/ml. Fluorescence intensity of supernatant fluid at 1) λ_{ex} = 280 nm, λ_{em} = 340 nm

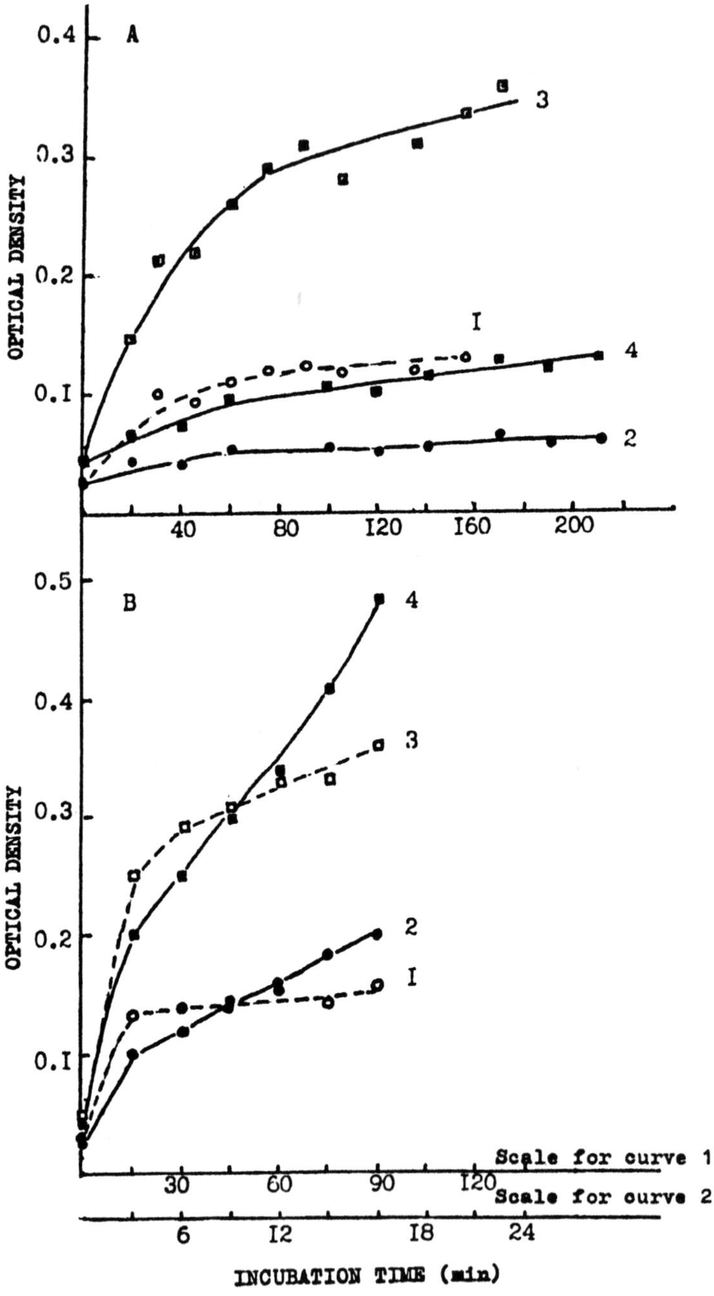

Figure 4 UV-absorbing compound release from E. coli M-17 during incubation in physiological solution at A) 48°C (1), 42°C (2), $\lambda = 280$ nm, 48°C (3), 42°C (4), $\lambda = 260$ nm, $c = 5.0*10^9$ cells/ml; B) 48°C (1), 55°C (2), $\lambda = 280$ nm, 48°C (3), 55°C (4), $\lambda = 260$ nm, $c = 5.0*10^9$ cells/ml. Fluorescence intensity of supernatant fluid at 1) $\lambda_{ex} = 280$ nm, $\lambda_{em} = 340$ nm

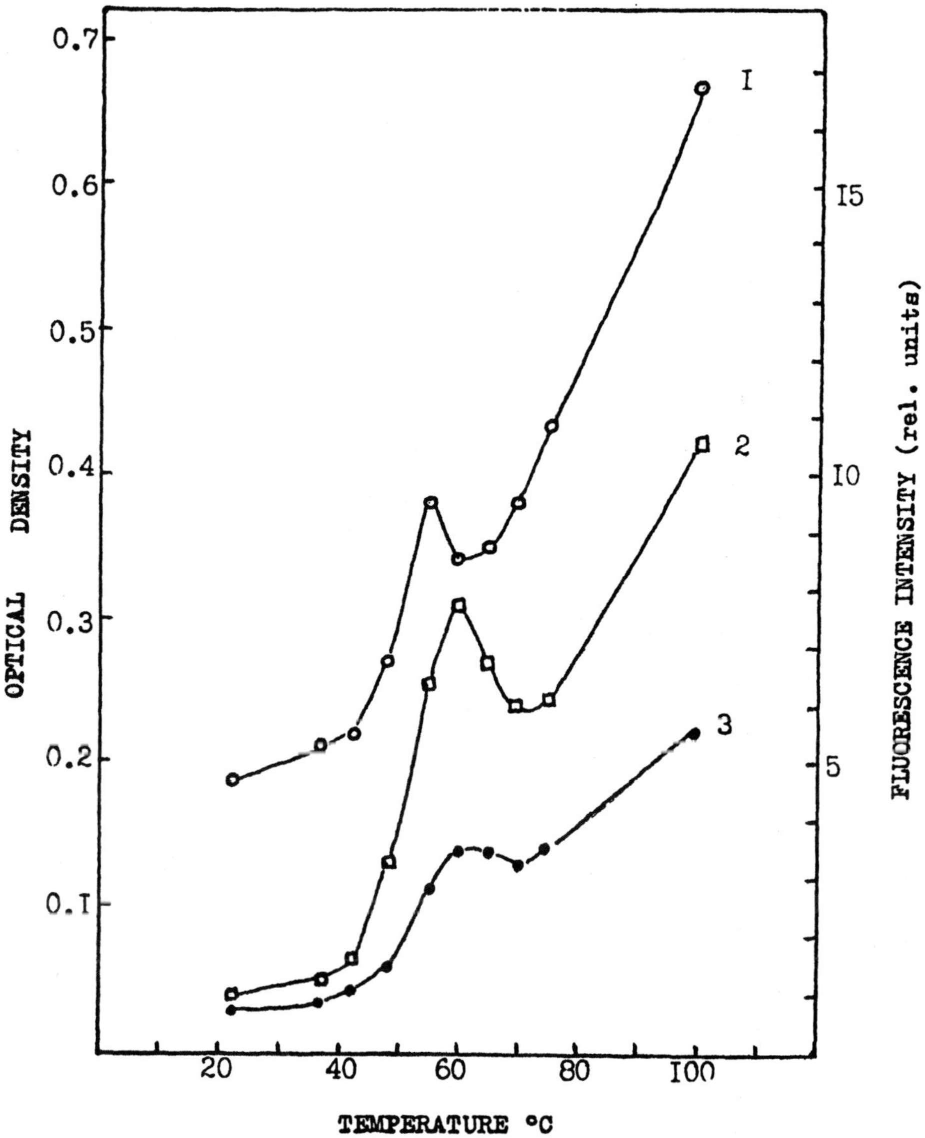

Figure 5 Release of fluorescent and UV-absorbing compounds from E. coli M-17 during incubation in physiological solution at different temperatures for 15 minutes. Fluorescence intensity of supernatant fluid at λ_{ex} = 280 nm, λ_{em} = 340 nm (1), optical density of supernatant fluid at 260 nm (2) and 280 nm (3).

of the cell envelope. The anomalous behavior of bacterial culture at about 60 C can be explained assuming that protein release is the active transport process and the optimal temperature of the transport system operation is below 60 C. The similar reasoning can be repeated for ultraviolet-absorbing compounds.

Physiological meaning of the described phenomenon is that protective reactions related to the metabolism inhibition occur in nature under stress conditions widely. These reactions can be carried out by the release of the cell metabolites necessary for the intensive metabolic process maintenance.

Conclusion

The protein release from Escherichia coli is a universal adaptive response of bacteria to stress. The observed phenomenon is unrelated to a well-known universal response, the synthesis of stressed proteins (unpublished results). The highly sensitive method used in this study for protein determination does not allow detailing these proteins. Different methods must be invoked for the solution of such problems. We believe that the described phenomenon will be applied in biotechnology for the production of biologically active substances (regulatory proteins, for example).

References

1. E.I. Melekhov, "Possible Principle of Damage Regulation and Defense Reaction of the Cell", *Zhurnal Obschei Biolgii*, Vol. 44, No. 3, 1983, pp. 386-397, Russian.

2. E.I. Melekhov, "The Regulation Principle of Cell Damage Rate and the Reaction of Protective Metabolism Inhibition", *Zhurnal Obshchei Biologii*, Vol. 46, No. 2, 1985, pp. 174-189, Russian.

3. C.W. Moss and M.L. Speck. "Biologically Active Peptide Release from Escherichia coli at Subzero Temperatures", *J. Bacteriol.*, Vol. 91, No. 3, 1966, pp. 201-207.

4. V.V. Lupashin, S.V. Kononova, Y.N. Ratner, A.V. Tsiomenko and I.S. Kulayev. "A New Secretory Protein of the Yeast S-cerevisiae Stimulated by Heat Shock", *DAN SSSR*, Vol. 317, No. 5, 1991, p. 1257-1260, Russian.

5. E.A. Burstein, N.S. Vedenkina and M.N. Ivkova. "Fluorescence and Tryptophan Residue Location in Protein Molecules", *Photochem. and Photobiol.*, Vol. 18, No. 4, 1973 pp. 263-279.

6. E.K. Roshchina, O.V. Dobrolez, and L.N. Petrov. "Spectrofluorimetric Analysis of Escherichia coli during the Process of Storage in Physiological Solution", *Microbiology*, Vol. 53, No. 6, 1984, pp. 833-837.

7. I.I. Zakharov, E.K. Roshchina, O.V. Dobrolez, L.M. Aleksin and L.N. Petrov. "Fluorescent Probes in Physiological Evaluation of Bacterial Population", "In: *Spectroscopy Methods in Physiology and Biochemistry*, (Spectroskopicheskie metody issledovaniia v phiziologii i biochimii), Leningrad: Nauka, 1987.

8. V. Ja.Munblit, V.L. Tal'roze and V.I. Trofimov. *Termoinactivatsiia mikroorganizmov*, (Heat Inactivation of Microorganisms), Moscow: Nauka, 1985, pp. 178-179.

Chromogenic Reaction 5-Aminosalicylic Acid: Unique Property and Identification Test of Bacteria Belonging to the Genus Klebsiella

E.P. Sivolodsky

Abstract

The taxonomic significance of chromogenic reaction with 5-aminosalicylic acid (5-ASA) on "Klebsiella 5-ASA" culture medium and its diagnostic characteristics was studied. Unique specific color reaction with 5-ASA has bccn found for medically significant Klebsiella species. A number of strains have color reaction with 5-ASA: 98,3 0, 7% of K. pneumoniae subsp. pneumoniae strains (n=290); 100% of K. pneumoniae subsp. rhinoscleromatis (n=34); 94, 7 3, 8% of K. pneumoniae subsp. ozaenae (n=38); 94, 5 3, 1% of Kl mobilis (n=55); 90, 8 2, 6% of K. oxytoca (n=120). K. planticola and K. terrigena have no color reaction with 5-ASA. All other species of bacteria (65 species of 28 genera) have a negative reaction. The presence of color reaction with 5-ASA confirms assignment of K. mobilis to the genus Klebsiella. According to the clinical diagnostic studies the culture medium "Klebsiella 5-ASA" makes possible one step recovering and identifying Klebsiella genus simultaneously on color test with 92, 2 2, 7% sensitivity and 100% specificity.

1. Introduction

Despite a century studying of Klebsiella, the taxonomic status of some species in this genus is still under discussion. In 1971 it was proposed to include Enterobacter aerogenes named Klebsiella mobilis to the genus Klebsiella [1]. The assignment of K. mobilis to the genus Klebsiella is confirmed by DNA-DNA hybridization method [2,3]. However, according to Bergey's Manual of Systematic Bacteriology K. mobilis species was not included in the genus Klebsiella [4]. In 1985 the property of bacteria specific only for the representative of the genus Klebsiella was found. This property is color reaction with 5-aminosalicylic acid (5-ASA) on a culture medium [5,6].

The purpose of the present work is to study taxonomic significance of Klebsiella property to show color reaction with 5-ASA and diagnostic characteristics of the culture medium "Klebsiella 5-ASA".

2. Methods

943 bacteria strains belonging to 70 species of 29 genera that include 553 strains assigned to all Klebsiella species have been studied (table). Typical strains were obtained from the collection of the L.A. Tarasevich GISK. The majority of K. pneumoniae subsp. ozaenae and K. pneumoniae subsp. rhinoscleromatis strains were from the Minsk Medical Institute collection; K. terrigena and some K. planticola strains were from that of the Russian Scientific Research Institute of Agricultural Microbiology. Other ones were recovered from clinical material. The following species of other bacteria (1-10 strains of each species) were under investigation: Alcaligenes faecalis, A. Denitrificans; Aeromonas hydrophilia, A. Caviae, A. sobria; Acinetobacter calcoaceticus, A. baumanii, A. lwoffii; Vibrio cholerae "non 01", V. metschnikovii, V. parahaemolyticus, V. alignolyticus, V. fluvialia I, V. fluvialis II, V. vulnificus; Plesiomonas shigelloides, Flavobacterium meningosepticus, Bordetella bronchiseptica, Sewanella putrefaciens, Escherichia coli; Shigella dystenteriae, S. Flexneri, S. boydii, S. sonnei; Citrobacter freundii, C. diversus, Edwardsiella tarda; Salmonella typhi, S. paratyphi B, S. typhimurium; Enterobacter cloacae, E. agglomerans, E. sakasaki, E. intermedium; Hafnia alvei,; Serratia marcescens; Proteus mirabilis, P. vilgaris; Providencia alcalifaciens, P. stuartii, P. rettgeri; Morganella morganii; Yersinia pseudotuberculosis, Y. enterocolitica; Erwinia

cartovora, E. herbicola; Micrococcus luteus, Enterococcus faecalis, Staphulococcus aureus; Bacillus subtilis, B. cereus; Pseudomonas aeroginosa, P. putida, P. florescens, P. aureofaciens, P. chloroaphis, P. mendocina, P. stutzeri, P. alcaligenes, P. pseudonalcaligenes, P. diminuta; Comamonas testosteroni, C. acidovorans; Xanthomonas maltophilia. Bacteria identification was performed on Bergey's Manual of Systematic Bacteriology [7].

In the research 5-ASA (Shostkinsk Chemical Plant or Sigma, USA), L-Arabinose (Reanal, Hungary), Histamine and D-Melezitose (Sigman, USA) were used. Nutrient agar medium was developed from fish hydrolysate.

The Klebsiella property to show color reaction with 5-ASA was studied on the solid culture medium "Klebsiella 5-ASA" developed by us [5,6]. The culture medium formula (g/L) is as follows: dehydrated nutrient agar 32.0-35.0; L-Arabinose 8.0-12.0; brom thymol blue 0.06-0.08; 5-ASA 4.5-5.5; distilled water 11; 1N NaOH for final pH 6.8-7.2 adjustment. To prepare the culture medium 32-35 g of nutrient agar id dissolved in 11 of distilled water, heated with agitation and sterilized by autoclaving at 120 C for 20 minutes. Then 1 g of 5-ASA, 2 g of L-Arabinose and 0.8 ml of 1.6% brom thymol blue alcohol solution is added to 200 ml of hot nutrient agar, pH 6.8-7.2 adjusted by adding 1 N NaOH solution until green color appears (about 6 ml) and the medium dispensed in sterile agar plates. The final product for performance is green and transparent. It can be used for 5 days and must be stored at 4-8 C.

To carry out the test a loopful of agar culture is inoculated like plaque (diameter 5 mm) on the surface of the "Klebsiella 5-ASA" medium segment, and incubated at 37 C for 18-24h. A positive result is a zone of dark brown color around the grown macrocolony. The Lawn of growth and the underlying medium are yellow or green. The clinical material (urine, feces, sputum, etc.) is inoculated by a loop or cotton swab on the agar surface and incubated at 37 for 18-24h. The identification of Klebsiella genus is carried out simultaneously with recovering if there is a dark brown zone around the grown colonies. The control identification of the isolated bacteria was conducted in accordance with Bergey's Manual of Systematic Bacteriology [7].

3. RESULTS AND DISCUSSION

It has been found that color reaction with 5-ASA is shown by the bacteria of the most prevalent and medically significant Klebsiella species. K. pneumoniae subsp. pneumoniae - 98, 3 0, 7%; K. pneumoniae subsp. ozaenae - 94, 7 3, 8%; K. pneumoniae subsp. rhinoscleromatis - 100%; K. oxytoca - 90, 8 2, 6%; K. mobilis - 94, 5 3, 1%. Rarely occurring in man, K. planticola and K. terrigena bacteria, as well as all of the studies strains of other ones (65 species, 28 genera), did not show any color reaction with 5-ASA (table). Thus, important taxonomic significance of color reaction was revealed as a unique specific property of Klebsiella bacteria genus except for K. terrigena and K.planticola species. The presence of this property in K. mobilis is of particular importance in that its taxonomic status is questionable. This fact shows that K. mobilis belongs to the genus Klebsiella.

Color reaction on culture medium is due to Klebsiella interaction with 5-ASA, where the presence of L-Arabinose is not necessary. the function of L-Arabinose is to stabilize the medium color. Brom thymol blue is necessary for visual correction of pH while preparing the medium. Any nutrient agar providing the growth of Klebsiella can be used as a nutrient basis for the culture medium. Color reaction with 5-ASA occurred in Klebsiella at 22-42 C (most intensively at 35-37 C) only under aerobic conditions. It is not observed in liquid culture media. The culture medium "Klebsiella 5-ASA' is highly sensitive to K. mobilis, X. pneumoniae subsp. pneumoniae, K. oxytoca (2-9 Cfu) like the sensitivity of original nutrient agar without supplements. It allows using this medium as a primary one for recovering and identifying Klebsiella species simultaneously.

The medium has low sensitivity for K. pneumoniae subsp. ozaenae and K. pneumoniae subsp. rhinoscleromatis - 10^5 - 10^6 Cfu. Therefore it can be used only for identifying the cultures of these species with heavy inoculation. The medium sensitivity for K. planticola and K. terrigena is 2-9 Cfu. Gram-position bacteria, Yersinia pseudotuberculosis, bacteria of genus Pseudomonas, Comamonas, Alcaligenes, inoculated in 10^6 - 10^7 Cfu do not grow due to the inhibitory properties of the medium.

The mechanism of color reaction with 5-ASA has also been established by us. It will be reported in the additional communication.

The culture medium "Klebsiella 5-ASA" was evaluated as a primary inoculation medium in current work of diagnostic laboratory. Samples had been streaked over the surface of agar plate quarter with "Klebsiella 5-ASA" medium and both on Endo and blood agar plates. The culture of 680 samples of urine and wound discharged specimens yielded Klebsiella growth in 103 samples. Klebsiella were recovered on "Klebsiella 5-ASA" medium from 95 samples out of 103 ones (sensitivity 92, 2 2, 7%) and simultaneously identified with positive color reaction in 24 hours after primary inoculation. Control identification of all strains showing color reaction with 5-ASA revealed their assignment to Klebsiella genus (specificity 100%). Among the isolated Klebsiella without color reaction were 4 strains of K. planticola detected with additional test of D-Melezitose fermentation, histamine utilization, gas production from lactose at 44, 5 C. As a whole K. pneumoniae subsp. pneumoniae were isolated from 86 samples, K. oxytoca from 8, K. mobilis from 5, and K. planticola from 4 ones. The culture medium inhibited the growth of all bacteria other than Klebsiella in 32, 2 1, 8% of samples. From other specimens bacteria association was isolated.

4. CONCLUSION

1. Color reaction with 5-aminosalicylic acid on "Klebsiella 5-ASA" medium is a unique specific property for Klebsiella species having medical significance. This property is manifested by K. pneumoniae subsp. pneumoniae (98, 3 0, 7%), K. pneumoniae subsp. ozaenae (94, 7 3, 8%), K. pneumoniae subsp. rhinoscleromatis (100%), K. oxytoca (90, 8 2, 6%), K. mobilis (94, 5 3, 1%), K. planticola and K. terrigena have no color reaction with 5-ASA. Investigated bacteria other than Klebsiella show a negative reaction.

2. The presence of color reaction with 5-ASA property in K. mobilis confirms assignment of this species to Klebsiella genus.

3. In laboratory practice "Klebsiella 5-ASA" culture medium enables one step isolation and simultaneous identification of Klebsiella with color test (sensitivity 92, 2 2, 7%, specificity 100%). Strains of Klebsiellae without color reaction require special testing for assignment of K. planticola and K. terrigena.

Table 16.1 Characteristerics of Klebsiella species and other bacteria on colour reaction with 5-ASA in "Klebsiella 5-ASA" culture medium

Bacteria Species	Number of strains studied	Strain origin	Number of strains showing colour reaction with 5-ASA	
			n	M ± m%
K. pneumoniae subsp. pneumoniae	290	ATCC 13886, NCTC 9127, other strains from clinical specimens	285	98, 3±0, 7
K. pneumoniae subsp. ozaenae	38	ATCC 11296, other strains from clinical specimens	36	94, 7±3, 8
K. pneumoniae subsp. Rhino-scleromatis	34	ATCC 13884, other strains from clinical specimens	34	100
K. mobilis	55	from clinical specimens	52	94, 5±3, 1
K. oxytoca	120	from clinical specimens	109	90, 8±2, 6
K. planticela	14	5 from RSRI of agricultural microbiology other strains from clinical specimens	0	0
K. terrigena	2	from RSRI of agricultural microbiology	0	0
Other species (65 species 28 genera)	390	from clinical specimens and environment	0	0

5. References

1. S. Bascomb, S.L. Lapage, W.R. Willcox, N.A. Curtis. "Numeral Classification of the Tribe Klebsiella" *J. Gen. Microbiol.*, Vol. 66, 1971, pp. 279-295.

2. D. Izard, F. Gavivi, P.A. Trinel, F. Krubwa, H. Leclerc. "Contribution of DNA-DNA Hybridization to the Transfer of Enterobacter Aerogenes to the Genus Klebsiella as K. mobilis". *Zbl. Bakt. Hyg. I. Abt. Orig. Cl.*, Vol. 257, 1980, pp. 257-263.

3. C. Ferragut, K. Kersters, J. De Ley. "Protein electrophoretic and DNA Homology Analysis of Klebsiella Strains" *System. Appl. Microbiol.*, N. 11, 1989, pp. 121-127.

4. I. Orskov. "Genus V. Klebsiella Trevisan 1885". In: *Bergey's Manual of Systematic Bacteriology* (N.R. Krig and J.C. Rolt Ed.) - Baltimore: Williams & Wilkins, Vol. 1, 1984, pp. 461-465.

5. E.P. Sivolodsky. A.S. 1313874 USSR. *Otkritija,* 1987, N. 20, p. 106 (in Russian).

6. E.P. Sivolodsky. "Specific Property of Bacteria Belonging to the Genus Klebsiella: Color Reaction with 5-aminosalicylic Acid in a Culture Medium". *Zhurn microbiol., epidemiol. i immunol.*, N. 12, 1988, pp. 26-29 (in Russian).

7. *Bergey's Manual of Systematic Bacteriology* (N.R. Krig and J.C. Holt Ed.) - Baltimore: Williams & Wilkins, Vol. 1, 1984, 964 pp.

TEMPERATURE EFFECT ON ANTIMICROBIAL SUSCEPTIBILITY OF BACTERIA INSIDE THE COLONIES

E.N. Shmidt, V.V. Tetz

ABSTRACT

The effect of physiological variations in temperature on enterobacteria antimicrobial susceptibility was studied. Marked differences in bacteria survival within the colonies in the presence of antimicrobial agents were registered. Maximal killing activity was observed at fabric temperature (42 C). This phenomenon may be associated with the physiological properties of bacterial colonies as the whole integral and isolated systems.

Key words: temperature; antimicrobial susceptibility; bacterial colonies.

INTRODUCTION

The effect of different environmental factors on bacteria is an important problem in microbiology, closely connected with bacteria survival in the presence of antimicrobial agents. As is known bacteria form colonies in natural conditions. There is evidence that bacterial cells within the colonies possess some properties specific to multicellular organism [1]. The colonies of Gram-negative and Gram-positive bacteria are separated from their external environment by surface films the basic part of which is represented by an elementary membrane covered with amorphous layers [2,3,4]. The cells of 3 to 5 superficial layers form capsular-like envelope similar to glycocalyx, described in some bacteria [1]. These surface structures may affect the penetration of different substances into the colonies.

It was shown earlier that in vivo bacteria also produce biofilms in which microorganisms are protected from some immune factors and antibiotics [5,6].

Temperature is an important environmental factor affecting physiology and virulence of pathogenic bacteria. Physiologic variations of the temperature in infectious diseases may cause a direct effect on different properties of the pathogens, including bacterial antimicrobial susceptibility. A few studies dealing with this problem revealed no consistent pattern of association between the temperature and the minimal inhibitory concentrations (MIC) or the rate of bacterial killing by antibiotics [7,8,9]. However, nothing is known about the temperature effect on the antibiotic activity against adherent bacteria within the colonies or biofilms.

In this study the effect of physiologic variations in temperature on clinically important enteric bacteria survival inside the colonies in the presence of antibiotics was examined.

MATERIALS AND METHODS

Six bacterial strains isolated from patients in the hospitals of St. Petersburg were studies: Shigella flexneri VT49, Shigella sonnei VT92, Escherichia coli VT1240, and three strains of uropathogenic Escherichia coli serotype 018ac. They possessed typical morphology, as well as tinctorial, biochemical, and serological characteristics of corresponding species and serogroups.

The antimicrobial agents, such as ampicillin (Amp), chloramphenicol (Chm), gentamicin (Gm), nalidixic acid (Nal), rifampicin (Rp), and tetracycline (Tc) were used for this study.

MICs were determined by the microdilution procedure described by Mackowiak et. al. [8], using overnight cultures adjusted to the concentrations of approximately $5x10^5$ colon forming units (CFU) per ml.

To examine the survival of bacteria in broth, overnight cultures grown at 37 C were diluted in aminopeptide broth, containing 2xMIC of antibiotic. After 2-hour incubation at 20-21 C (room temperature), 37 C and 42 C, the number of CFU per ml. was determined. The bacterial colonies were grown at 37 C in Petri dishes containing 25.0 ml. of agar. The pieces of the agar (70x40mm) with several 1-day-old colonies were cut out, placed in the Petri dishes with the agar containing antimicrobial agent and incubated at an appropriate temperature for 24 hours. An adequate amount of each antibiotic was used so that its concentration in the upper layer of the agar after diffusion was equal to 10-50xMICs. After incubation ;the CFU of 5 colonies with similar sizes were determined. The survival of bacteria is presented in the percentage of CFU, taking the CFU number in the intact 1-day-old colony as 100%.

The colony size was determined as an optical density of the suspension of 5 colonies in the constant volume of 0.15M NaCl, measured nephelometrically at the wave length of 590 nm (OD590).

All data are presented as an average of 5 independent experiments with standard deviations.

RESULTS

The effect of variations in the incubation temperature on the MICs of various antimicrobial agents for the test bacteria were determined at the incubation temperatures of 37 C, 42 C, and room temperature (Tab. 1). There was no greater difference than one twofold dilution between the MICs. The MICs were lower at room temperature than at 37 C in most cases. The MICs at 37 C and 42 C were not equal only for Gm.

The effect of variations in the incubation temperature on antibiotic bactericidal activity was more pronounced. The survival of planktonic bacteria (Tab. 2) after 2-hour incubation in the presence of 2xMIC of antimicrobial agents was 10-100 times higher at room temperature and differed at 37 C and 42 C insignificantly. All variations in the survival were eliminated after 24 hours: more than 99.99% of the bacteria were unviable at each temperature.

Temperature-dependent variations in the bacterial survival after 24-hour incubation were retained when the bacteria were inside the colonies (Tab. 3). All microorganisms within 1-day-old colonies became essentially resistant to all tested antibiotics. 10-50xMICs of the drugs were used to examine the killing effect. Gm showed the maximal activity against the bacteria inside the colonies (10xMICs), while NaCl revealed the minimal one (50xMIC).

The survival of bacteria like the broth cultures increased at room temperature for all used antibiotics. Marked differences in the killing activity were also observed at 37 C and 42 C. In 14 (67%) out of 21 combinations of tested bacteria and antibiotics the survival decreased more than fourfold at 42

C, and in 4 cases (20%) it was 10 times less than at 37 C. Antibiotic susceptible patterns were not altered when the colony formation took place at the temperature of subsequent killing activity determination.

The bacteria growth was affected by variations in temperature within the experimental range. This relationship was evaluated in the studies with two strains: E. coli VT1240 and S. Sonnei VT92. In control broth cultures without any contacts with the antibiotic (Fig. 1) after 2-hour incubation the CFU number increased only twofold at room temperature, whereas at 37 C and 42 C it went up about 10 times. After 24-hour incubation maximal viable count was determined at 37 C. At room temperature it was rather low, and it was more than 2 times less at 42 C. The CFU number increased about 20-50 times.

The determination of the bacterial survival in intact colonies (Fig. 2) showed no more than fivefold increase in the viable count. It was maximal at room temperature, and minimal (about twofold less) at 42 C. The colon size of E. coli was maximal at room temperature, while for S. sonnei it was maximal at 42 C.

Discussion

In the present study we have examined the relationship between the temperature and bactericidal antibiotic activity against the bacterial colonies. We have shown marked differences in the clinically important bacteria survival within the colonies at the temperature of external environment and normal human temperature, as well as at normal and fabric human body temperatures. In accordance with the data or other investigations [7,8] our experiments have revealed that the bacterial survival in the presence of antibiotics could be decreased, remain unaffected or slightly augmented by the increase in incubation temperature, depending on the examined antibiotic and bacterial strains. However, the increased temperature usually (87% of all combinations tested) caused more or less pronounced decrease in the bacteria survival within the colonies. These data are of practical importance in that human pathogens, including enteric and uropathogenic bacteria, form biofilms and microcolonies during infection [5]. It should be mentioned that the temperature effect on the bacterial survival, demonstrated in this study, would not be reflected as a result of the tested conventional susceptibility, since the activities of antimicrobial drugs in clinical laboratories are determined as MICs or MBCs, using bacterial cultures in broth or on agar outside the colonies or biofilms. Under these conditions the effect of temperature is minimal. Nevertheless, the temperature-dependent differences for bacterial survival in the presence of antimicrobial agents, observed in this study, may become fairly significant in patients.

The mechanisms responsible for the effect of temperature on antibiotic susceptibility are unknown. Numerous alterations in bacterial structure and physiology have been observed due to the temperature changes. Microorganism cultivation at different temperatures caused the changes in the fatty acid composition of cytoplasmic and outer membranes [10,1], the LPS content [10,11,12,13], the outer membrane protein composition [11,14], the activity of membrane-associated enzymes and transport systems [15,16]. All these structural and functional characteristics may affect antibiotic penetration into microbial cells [17]. The relationship between the temperature and the growth rate of bacteria is well known [18] and the correlation between the growth rate and bactericidal activity of some antimicrobial agents has been established [19,20,21]. Out data show that comparably small differences in the growth rate of bacteria within the colonies are accompanied by pronounced survival variations in the presence of antibiotics. The metabolic activity of bacteria also depends on the cultivation temperature [18]. It has been demonstrated that the rate of protein turnover increases with the temperature [22]. Physiologic variations in the temperature also alter the level of bacterial DNA supercoiling [23], subsequently affecting numerous genes expression [24]. The above mentioned parameters may influence the bactericidal activity of antimicrobial agents with the corresponding mode of action.

On the other hand, the bacterial colonies as the whole integral and isolated systems possess unique properties, which in their turn may vary with the temperature alterations. The metabolic activity of the individual cells decreases in colonies [25]. Moreover, the bacterial colonies form additional barriers for the penetration of different substances (surface film and capsule-like envelopes in the cells of superficial layers)[1,2,3]. This is the evidence of the alternations in morphology, the rate of colony formation and aging at an increased temperature [4]. Electromicroscopic analysis of bacteria ultrastructure within the colonies showed that the bacterial colonies after cultivation at room temperature or 37 C contained typical morphological cells, whereas at 42 C a great number of cells were at different stages of destruction, many of them containing noncrystalline granules. It is worth noting that the contacts between the cells in the colonies at a higher temperature were damaged. The formation of a surface film began earlier at 42 C than at lower

Table 1. Minimal inhibitory concentrations of antibiotics for tested bacteria at room temperature, 37°C and 42°C

Bacterium	t°C	MIC* Antibiotic				
		T_c	Chm	Gm	Rp	Nal
Shigella flexneri VT49	R**	0.25	0.5	2.0	4.0	2.0
	37°C	0.5	0.5	2.0	8.0	2.0
	42°C	0.5	0.5	1.0	8.0	2.0
Shigella sonnei VT92	R	0.25	0.5	2.0	4.0	2.0
	37°C	0.5	0.5	2.0	8.0	2.0
	42°C	0.5	0.5	1.0	8.0	2.0
Escherichia coli VT1240	R	1.0	1.0	1.0	8.0	1.0
	37°C	1.0	2.0	2.0	8.0	1.0
	42°C	1.0	2.0	1.0	8.0	2.0

*Values expressed in mkg of antibiotics per ml
**Room temperature

Table 2. Survival of the tested bacteria in broth after 2.5 hours in the presence of antibiotics at room temperature, 37°C and 42°C

Bacterium	t°C	CFU/ml Antibiotic	
		Amp	Nal
Shigella sonnei VT92	R*	$(2.0\pm0.3)x10^6$	$(1.4\pm0.1)x10^7$
	37°C	$(3.4\pm0.3)x10^4$	$(4.7\pm1.1)x10^5$
	42°C	$(2.4\pm0.4)x10^4$	$(2.8\pm0.2)x10^5$
Escherichia coli VT1240	R	$(4.5\pm0.5)x10^6$	$(1.2\pm0.3)x10^6$
	37°C	$(4.2\pm0.6)x10^4$	$(1.5\pm0.3)x10^6$
	42°C	$(1.4\pm0.1)x10^4$	$(1.9\pm0.3)x10^6$

*Room temperature

Table 3. Survival of the tested bacteria inside colonies in the presence of antibiotics (percentage of contact 1-day-old colony CFU) at room temperature, 37°C and 42°C

Bacterium	t°C	MIC*			
		Antibiotic			
		Amp	Chm	Gm	Rp
Shigella flexneri VT49	R**	-**	-	-	-
	37°C	19.3±2.0	23.5±0.5	4.0±0.8	19.1±3.9
	42°C	2.8±0.7	16.0±1.0	0.42±0.12	5.1±1.5
Shigella sonnei VT92	R	-	-	-	43.8±2.7
	37°C	14.0±2.9	22.0±0.5	4.6±2.5	14.2±0.1
	42°C	1.6±0.4	14.5±0.5	0.32±0.19	3.3±0.7
Escherichia coli VT1240	R	-	-	-	87.2±5.4
	37°C	12.5±0.5	-	-	16.3±1.5
	42°C	5.9±0.9	-	-	14.7±4.7
Escherichia coli 018ac/1	R	40.1±8.0	25.3±2.1	45.7±9.2	60.4±8.2
	37°C	6.6±1.0	5.2±0.9	0.27±0.07	7.9±2.1
	42°C	0.91±0.2	0.32±0.08	0.04±0.02	7.1±1.2
Escherichia coli 018ac/2	R	51.7±11.0	29.2±4.5	42.4±9.1	-
	37°C	5.4±1.0	1.0±0.23	0.04±0.02	9.0±2.9
	42°C	4.0±1.2	0.18±0.05	<0.001	13.9±2.3
Escherichia coli 018ac/3	R	44.8±2.5	41.1±6.1	52.6±4.1	-
	37°C	5.0±0.4	8.6±2.0	0.28±0.09	-
	42°C	0.86±0.12	0.78±0.17	0.14±0.06	-

*room temperature
**not determined

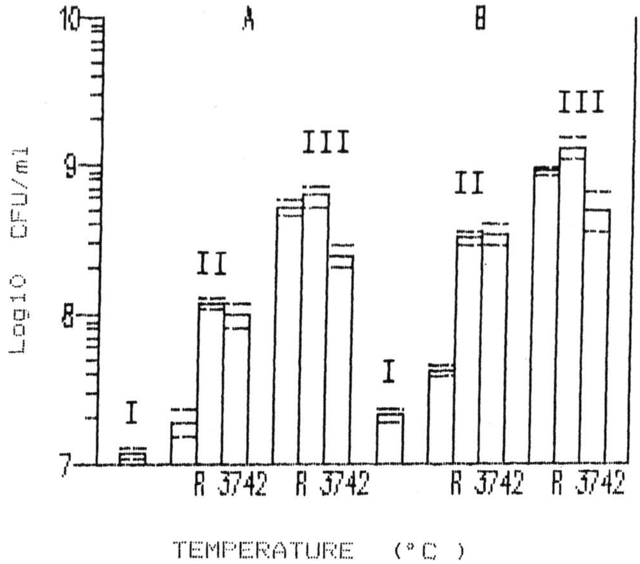

Figure 1 Growth of tested bacteria in the broth culture at different temperature. A - S. sonnei, B - e. coli. Bar graph illustrates CFU/ml after I - 0 hour, II - 2 hours, III - 24 hours of incubation at the appropriate temperature. R-room temperature.

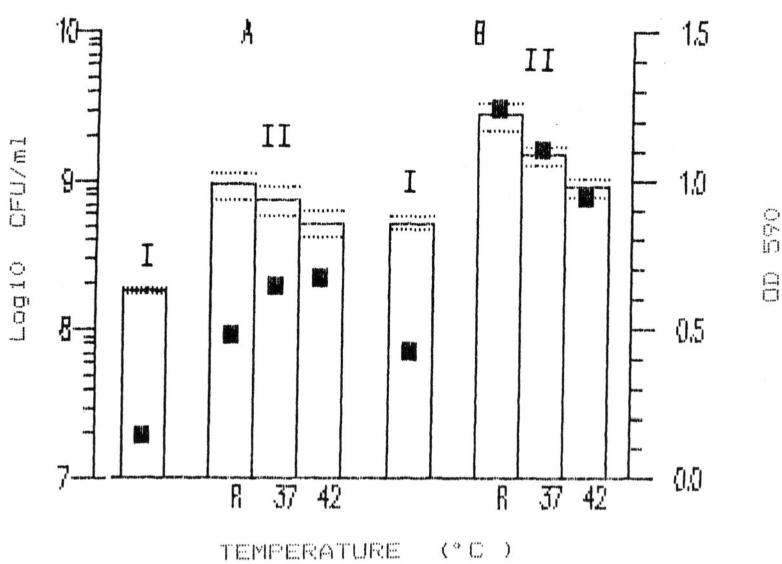

Figure 2 Growth of tested bacteria within the colonies at different temperatures. A - S. sonnei, B - e. coli.
I - 1-day-old colony, II - 2-day-old colony. Bar graph illustrates CFU number in one colony, - colony size (0D590)

temperatures. Only a few cells in broth culture were damaged or contained noncrystalline granules after cultivation at 42 C [4]. It should be mentioned that temperature-dependent variations in these properties may account for the observed differences in the bacteria survival within the colonies in the presence of antimicrobial agents in comparison with planktonic bacteria in the broth cultures.

Conclusion

1. Physiologic variations in incubation temperature alter MIC values and the survival of broth bacterial cultures in the presence of antimicrobial agents insignificantly.

2. The bacteria within the colonies become essentially resistant to all antibiotics tested.

3. Marked differences in the bacteria survival within the colonies in the presence of antimicrobial agents were observed at the temperatures tested, which cannot be associated with the growth rates of microorganisms directly.

4. Temperature-dependent variations in the survival differ with the antibiotic and bacterial strain tested. However, in most cases the survival was maximal at room temperature and minimal at 42 C. This maybe associated with the properties of bacterial colonies as integral systems. This phenomenon is of a clinical importance in the sense that pathogenic bacteria form colonies during infection, and hyperthermia is a common feature of infectious diseases.

References

1. V.V. Tetz, O.V. Rybalchenko, and G.A. Savkova. "Ultra-structural Features of Microbial Colony Organization." *J. Basic Microbiol.* Vol. 30, No. 8, 1990, pp. 597-607.

2. V.V. Tetz, O.V. Rybalchenko, and G..A. Savkova. "Surface Films of Escherichia Coli Colonies." *FEMS Microbiol.* Lett. Vol. 107, 1993, pp. 231-240.

3. V.V. Tetz, O.V. Rybalchenko, and G.A. Savkova. "The Ultrastructure of the Surface Film in Bacterial Colonies." *J. Gen. Microbiol.* Vol. 137, 1993, pp. 1081-1086.

4. V.V. Tetz, O.V. Rybalchenko, and G.A. Savkova. "Morphophysiologic Characteristics of Escherichia Coli and Shigella Flexneri 2a carrying the *htp*R15 Gene." *J. Basic Microbiol.* Vol. 33, No. 2, 1993, pp. 131-139.

5. J.W. Costerton. "Bacterial Biofilms in Nature and Disease." *Ann. Rev. Microbiol.* Vol. 41, 1987, pp. 435-464.

6. I. Ketyi. "Escherichia Coli Resistance to Some Antibiotics and biocides in the Intestinal Biofilm." *Acta Microbiol. Hung.* Vol. 38, No. 1, 1991, pp. 33-41.

7. P.A. Mackowiak, A.E. Ruderman, R.M. Martin, W.J. Many, J.W. Smith, J.P. Luby. "Effects of Physiologic Variations in Temperature on the Rate of Antibiotic-induced Bacterial Killing." *Am. J. Clin. Pathol.* Vol. 76, 1981, pp. 57-62.

8. P.A. Mackowiak, M. Marling-Cason, R.L. Cohen. "Effect of Temperature on Antimicrobial Susceptibility of Bacteria." *J. Infect. Dis.* Vol. 145, No. 4, 1982, pp. 550-553.

9. P.E. Wheat, T.G. Winstanley, R.C. Spencer. "Effect of Temperature on Antimicrobial Susceptibility of Pseudomonas Maltophila." *J. Clin. Pathol.* Vol. 38, No. 9, 1985, pp. 1055-1057.

10. S. Rottem, O. Markovitz, S. Razin. "Thermal Regulation of Lipopolysaccharides and Phospholipids of Proteus Mirabilis." *Europ. J. Biochem.* Vol. 85, 1978, pp. 445-450.

11. A.M. Kropinski, V. Lewis, D. Berry. "Effect of Growth Temperature on Lipids, Outer Membrane Proteins, and Lipopolysaccharides of Pseudomonas Aeruginosa." *J. Bacteriol.* Vol. 169, No. 5, 1987, pp. 1960-1966.

12. S. Schleht. "Effect of the Cultivation Temperature on the Production of Biomass and Lipopolysaccharide as well as the Chemical of S-form and R-mutants of the Genus Salmonella." *Zbl. Bakteriol. Mikrobiol. Hyg. [A]* Vol. 261, No. 2, 1986, pp. 147-160.

13. K. Wartenberg, W. Knapp, N.M. Ahamed, C. Widemann, H. Mayer. "Temperature-dependent Changes in the Sugar and Fatty Acid Composition of Lipopolysaccharides from Yersinia Enterocolitica Strains." *Zbl. Bakteriol. Mikrobiol. Hyg. I.* Vol. 253, No. 4, 1983, pp. 523-530.

14. B. Lugtenberg, R. Peters, H. Bernhiemer, W. Berendsen. "Influence of Cultural Conditions and Mutations on the Composition of the Outer Membrane Proteins of Escherichia Coli." *Mol. Gen. Genet.* Vol. 147, No. 3, 1976, pp. 251-262.

15. H. Sandermann. "Regulation of Membrane Enzymes by Lipids." *Biochem. Biophys. Acta,* Vol. 515, No. 3, 1978, pp. 209-237.

16. M.O. Eze, R.N. McElhaney. "Effect of Temperature on Active Amino Acid Transport in Escherichia Coli Strain 7." *Microbios.* Vol. 58, No. 236-237, 1989, pp. 173-182.

17. H. Nikaido. "Outer Membrane Barrier as a Mechanism of Antimicrobial Resistance." *Antimicrob. Agents Chemother.* Vol. 33, No. 11, 1989, pp. 1831-1836.

18. J. Ingraham. "Effect of Temperature, pH, Water Activity, and Pressure on Growth." In: *Escherichia coli and Salmonella typhimurium: cellular and molecular biology.* F.C. Neidhardt (Ed.), Washington, D.C.: Am. Soc. Microbiol. Vol. 2, 1987, pp. 1543-1554.

19. E. Tuomanen, R. Cozens, W. Tosch, O. Zak, A. Tomasz. "The Rate of Killing of Escherichia Coli by Beta-lactam Antibiotics is Strictly Proportional to the Rate of Bacterial Growth." *J. Gen. Microbiol.* Vol. 132, 1986, pp. 1297-1304.

20. N.E. Wright, and P. Gilbert. "Influence of Specific Growth Rate and Nutrient Limitation on the Sensitivity of Escherichia Coli towards Polymyxin B." *J. Antimicrob. Chemother.* Vol. 20, 1987, pp. 303-312.

21. M.E. Muir, R.S. Van Heeswick, and B.J. Wallace. "Effect of Growth Rate on Streptomycin Accumulation by Escherichia Coli and Bacillus Megaterium." *J. Gen. Microbiol.* Vol. 130, 1984, pp. 2015-2022.

22. M. Strnadova, R. Prasad, H. Kucerova, J. Chaloupka. "Effect of Temperature on Growth Rate and Protein Turnover in Bacillus Megaterium." *J. Basic. Microbiol.* Vol. 26, No. 5, 1986, pp. 289-298.

23. E. Goldstein, K. Drlica. "Regulation of Bacterial DNA Supercoiling: Plasmid Linkage Numbers Vary with Growth Temperatures." *Proc. Natl. Acad. Sci.,* Vol. 81, No. 13, 1984, pp. 4046-4050.

24. G.J. Pruss, K. Drlica. "DNA Supercoiling and Prokaryotic Transcription." *Cell,* Vol. 56, No. 4, 1989, pp. 512-532.

25. S.J. Pirt. *Principles of Microbe and Cell Cultivations.* London: Blackwell. Sci. Publ. Oxford, 1975, 271 pp.

INDEX

A

B

D

E

F

G

H

M

N

O

P

R

S

T

U

V

W